Colusa County Free Library
738 Market Street
Colusa, CA 95932
Phone: 458-7671

CUP   PRINCETON BRANCH

THE CAT OWNER'S HANDBOOK

THE CAT OWNER'S HANDBOOK

Graham Meadows
and Elsa Flint

CAXTON EDITIONS

This edition published 2002 by Caxton Editions
an imprint of The Caxton Publishing Group

First published in the UK in 2001 by
New Holland (Publishers) Ltd
London • Cape Town • Sydney • Auckland

Copyright © 2001 New Holland Publishers (UK) Ltd
Copyright © 2001 in text: Graham Meadows and Elsa Flint
Copyright © 2001 in illustrations: New Holland Publishers (UK) Ltd

All rights reserved. No part of this publication may be reproduced, stored in a retrieval system or transmitted, in any form or by any means, electronic, mechanical or otherwise, without the prior written permission of the copyright holders and publishers.

ISBN 1-84067-3311   hardback

Publisher: Mariëlle Renssen
Managing Editors: Claudia Dos Santos, Mari Roberts
Managing Art Editor: Peter Bosman
Editor: Gail Jennings
Designer: Geraldine Cupido
Illustrations: Daniël van Vuuren
Picture researcher: Sonya Meyer
Proofreader and Indexer: Gudrun Kaiser
Production: Myrna Collins
Consultant: Bas Hagreis

Reproduction by Hirt & Carter (Cape) Pty Ltd

Printed and bound in Singapore by Tien Wah Press (Pte) Ltd

HALF TITLE PAGE: The hunting instinct of your pet cat lies just beneath its 'domesticated' exterior.
FULL TITLE PAGE, LEFT: Strange as this may seem, cats and pet rabbits can become good friends – your cat may even find its way into the rabbit hutch and laze in the sawdust with its rabbit companions.
FULL TITLE PAGE, RIGHT: Colourwise, mixed-breed littermates need not bear any resemblance to one another.

Publisher: Mariëlle Renssen
Managing Editors: Claudia Dos Santos, Mari Roberts
Managing Art Editor: Peter Bosman
Editor: Gail Jennings
Designer: Geraldine Cupido
Illustrations: Daniël van Vuuren
Picture researcher: Sonya Meyer
Proofreader and Indexer: Gudrun Kaiser
Production: Myrna Collins
Consultant: Bas Hagreis

Reproduction by UNIFOTO Pty Ltd
Printed and bound in Singapore by Tien Wah Press

Although the author and publishers have made every effort to ensure that the information contained in this book was correct at the time of going to press, they accept no responsibility for any loss, injury or inconvenience sustained by any person using this book.

ABOVE: Kittens are insatiably curious, and at six weeks are happy to begin interacting with strangers.
TOP: Abyssinian cats are not only elegant and charming – they are particularly intelligent and learn tricks easily.
OPPOSITE TOP: The father of these tabby kittens is a black cat, but black coat colour genes are recessive. This means that all tabby/black cross-breed kittens are tabby.

ABOVE: Kittens are born with an excellent ability to detect movement – not even this well-camouflaged green cricket is safe from the quick-witted kitten.
OPPOSITE TOP: Through their interaction with pets, children learn about love, death and respect for other living creatures.

CONTENTS

Cats and people	10
Your new cat	22
Caring for your cat	30
Nutrition	44
Understanding your cat	56
Breeding and reproduction	72
The mature cat	76
Protecting your cat's health	84
Monitoring your cat's health	96
First aid	110
Glossary	124
Index	126

CATS AND PEOPLE

Dogs have masters.
Cats have staff. *Anon.*

Those of us who are owned by cats may well subscribe to the theory that humans didn't domesticate the cat at all, but that the cat domesticated itself by walking into, adapting to and (in many cases) taking over people's lives. With few exceptions the modern domestic cat remains independent and solitary, and has an indefinable wild streak. It gives you a look that implies: 'I may live in your household, but don't expect me to conform.'

The origin of the domestic cat

A distant ancestor of today's domestic cat may have been Martelli's wild cat (*Felis lumensis*), a species now extinct. It was similar in size to today's small wild cats. About 600,000–900,000 years ago it may have given rise to *Felis silvestris*, from which three distinct types evolved according to the region and environment in which they lived. These were the central European or Forest wild cat (*F. silvestris*

ABOVE: Male lions (*Panthera leo*) within a pride can be particularly tolerant of cubs.
TOP: Despite their reputation for aloofness, it is not unusual for a cat to head-butt and lick its owner to show affection.

silvestris), the Asiatic desert cat (*F. silvestris ornata*) and the African wild cat (*F. silvestris lybica*). The latter inhabited most of Asia and North Africa, and because the process of domestication of the cat occurred mainly in the Middle East, the African wild cat was almost certainly the principal ancestor of modern domestic cats.

Domestication

For the cat, as for other domestic animals, the process of domestication occurred over a long period of time. Wild cats would have associated with humans once the latter stopped being hunter-gatherers and formed permanent settlements, grew grain crops and set up grain stores. Grain stores would have attracted mice and rats, which in turn would have attracted wild cats.

Any sensible agriculturist would quickly have seen the advantage of encouraging these cats to help control the vermin, so a loose but mutually beneficial association would have been forged.

Just when the process of domestication started is unclear, though, and our estimates rely on archaeological discoveries and the excavation of cat remains that can be shown to be closely associated with humans. Although various cat remains have been found in Egyptian archaeological sites dating to 6700BC, there is no firm evidence that these were domesticated animals, and they are more likely to have been wild cats. If you accept that finding a cat skeleton buried with a person is evidence that the cat was domesticated, then a 7000-year-old burial site at Mostagedda, in Egypt, is evidence enough. There, excavations revealed a man buried with two animals at his feet: a cat and a gazelle.

If this doesn't convince you, then you need to move forward 2500 years and to the earliest depiction of cats in Egyptian tomb art. Cat remains recovered from an archaeological site in the Indus Valley, dated at 2000BC, could well be from a domesticated variety, and paintings and inscriptions from the same period portray cats in situations that suggest that they were domesticated.

ABOVE: Many statues were made of the Egyptian cat goddess Bast. This bronze figure is dated between 664 and 525BC.

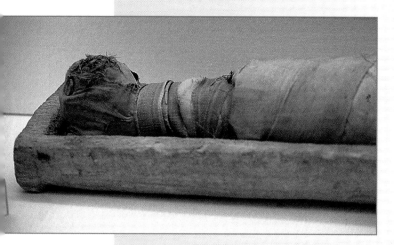

Cat worship and culture

Many thousands of years ago a cat cult was well established in ancient Egypt. There was a feline goddess, Mafdet, a snake-killer and protector of the pharaoh in the royal palace, whose pictures appear in magic formulas carved on pyramid chambers of the Fifth and Sixth Dynasties (before 2280BC).

The ancient Egyptians recognized the cat's role as a guardian of grain stores, protecting the animal by law and keeping sacred cats in their temples. In the temple of the cat goddess Bast or Pasht, from which the word 'puss' is said to have arisen, many thousands of cats were mummified and laid in tombs. Excavations at other sites have also revealed large numbers of mummified cats, and the height of the cat cult is thought to have occurred at around 500BC, when many other animals were also a subject of worship. It was once thought that all the mummies were of household cats that had died from natural causes and whose remains had been presented to the temple by their mourning owners. More recently researchers have concluded that many of them were cats specially bred for sacrifice, because they had died from a broken neck and many were merely kittens.

From that time on there is plenty of evidence to show that cats became well established in Egyptian homes. A painting at Thebes, in the tomb of the harbour-master May and his wife Tui (dated around 1600BC), portrays a ginger cat sitting beneath Tui's chair. It wears a collar, and its leash is tied to a chair leg. The inference is that it was a pet, although this could be disputed.

A picture in the tomb of someone named Baket (dated around 1500BC) depicts a house attendant watching a cat that is eyeing a rat. Other tombs in Thebes also contain paintings of cats. One of them, dated at 1400BC, depicts a kitten sitting on the lap of the sculptor Ipuy. There are also some interesting, though inconclusive, artifacts to suggest that by this period in history cats were not only kept as pets in homes, but also used to help people to hunt. At least three tomb paintings, one of them in the tomb of the sculptor Nebuman (around 1400BC), show cats apparently participating in the action while wildfowlers are using throwing sticks to catch and kill ducks and other birds. Were these cats helping to flush out game from the reed beds and/or helping to retrieve it? A sceptic might suggest that they were simply there to take advantage of a free lunch.

The taming of the cat

It has been suggested that during the process of domestication a genetic change to wild temperament (a 'domestication mutation') must have occurred to reduce the wild cat's innate aggression and make domestication possible. The basis for this reasoning is that in wild cats tameness (lack of

ABOVE: Don't be surprised to hear your mother cat purring loudly when she suckles her kittens.
TOP: Cats were worshipped in ancient Egypt, and many were mummified to accompany their owners into the next world.

CATS AND PEOPLE

Taming

aggression) is not inherited; although individual animals can be tamed, their kittens are born with a wild temperament and in their turn must also be tamed. In the domestic cat, kittens inherit tameness from their mother – therefore, the reasoning goes, some genetic change must have occurred in the domestic cat to cause this.

The idea of a domestication mutation is intriguing, for its exponents suggest that when this occurs it prevents the development of certain adult behaviour patterns, with the result that adult animals still retain some juvenile behaviours. Retaining these behaviours makes them better suited to domestication.

The term for this is neotony, and it functions as follows. In the wild, adult cats are solitary. A close-knit 'family' group is formed when a female gives birth to and rears her kittens, but once the kittens become independent there is no continuing association, and each individual becomes a 'loner'.

Domestic cats, on the other hand, behave rather differently. They are more gregarious, and the suggestion is that this is because they retain some of their 'kittenish' instinct to keep together. There are several examples to demonstrate this. If the owner of a female cat that has given birth to kittens decides to keep one or more of those kittens once they have been reared, the mother and offspring will often form close family bonds.

Even when domestic cats are feral, their families tend to stay together, while in urban areas, where there are comparatively dense domestic cat populations, unrelated adults will often form loose associations. Groups of them may even meet together at certain times of the day for 'cat conferences', which seem to be the cat equivalent to humans 'hanging out' together.

Neotony could arise from a genetic mutation, but it could also result from the process of human selection. People would choose to keep and breed the cats that were the easiest to manage. Those displaying juvenile characteristics were more family-oriented and less independent than adults, and therefore more suited to life within a human family.

Neotony is not just a characteristic of domestic cats. It occurs in domestic dogs, too, where adults retain certain puppy characteristics that make it easier to integrate them into a human family.

TOP: A lioness (*Panthera leo*) is forever vigilant, watching for potential danger to her cubs, even when they are no longer tiny and hopelessly vulnerable.

Whether such a genetic change occurred, and if so, when, we shall never know. We can only surmise that people kept the kittens of wild cats, and that some of these (probably females) proved tame enough to keep to adulthood and breed from. Eventually, for various reasons, kittens were born that were less aggressive and more suited to living with humans.

Nevertheless, the domestic cat's wild temperament is only just below the surface, and not all cats show the same degree of 'tameness'. There is a wide range of temperaments within the domestic cat population – some cats are extremely tame and others have a definite wild streak. Also, lack of aggression in domestic cats needs to be reinforced by human contact from an early age. If it is not, then some of the cat's wild attributes reappear. For example, kittens born to a domestic cat that has gone feral are distrustful of humans, and must be subjected to at least a basic taming process in order to adapt to living in a human home.

The spread of cats

As more trade routes developed between countries around the Mediterranean and in Asia, the spread of domestic cats also grew. Around 900BC, Phoenician traders took them to Italy, and from there they spread slowly across Europe, during which time genes from the European wild cat were introduced (by accident, design, or both).

We know that cats had arrived in England by AD1000, during the period of early Viking settlement, because cat remains have been discovered at several archaeological sites dating from that period (including the ancient Viking village of Jorvik, in York, England).

All these cats were shorthaired, but further to the east, longhaired varieties were being developed. It has been suggested that the gene for long hair may have come from the manul (*Felis manul*) of central Asia, but it is more likely to have originated from artificial selection for the gene that produces long hair. The gene for long hair spread from southern Russia into Pakistan, Turkey and Iran, and eventually showed up in the Angora and Persian breeds. Longhaired varieties arrived in Italy from Turkey during the 16th century, at about the same time as the Manx cat arrived in the Isle of Man, brought in from the Far East by Spanish traders who regularly plyed those routes.

The first colonists took shorthaired cats with them to the New World, and later settlers took a variety of cats to Australia and New Zealand. Domestic cats are now distributed throughout the world.

Coat colours and patterns

Some of the coat colours and patterns in domestic cats are thought to be very old, as they have had time to spread all over the world. They include black, blue (a slate-grey colour which is a 'dilute' form of black) and orange (ginger). The Siamese and Burmese colour patterns, on the other hand, are

ABOVE: Most, but not all, ginger cats are male. The coat colour is a sex-linked gene.
TOP: The origin of the Russian Blue is uncertain. It is said that sailors brought back specimens to Britain from the northern Russian port of Archangel.

CATS AND PEOPLE

Breeding

more recent and originated in southeast Asia – they were preserved and spread because of human interest.

Some colour patterns spread of their own accord. For example, some hundreds of years ago in Britain the blotched tabby appears to have arisen as a mutation of the striped tabby. For reasons yet to be fully explained, it appears that the blotched tabby and black cat are better able to thrive in a high-density urban environment, and in some areas (especially in England) these colours are becoming predominant among alley cats and non-pedigree domestic cats.

Breeding

By their very nature, cats, especially mature males (toms), are wanderers. Before the concept of selective breeding about 150 years ago, this wanderlust in domestic cats provided plenty of opportunity for the intermingling of genes. If there were two distinct races of cat in any region they blended over a period of time, so we cannot be sure of the origin of many of our modern domestic breeds.

Nevertheless, studies of the skeletal structure, body type and hair length of modern breeds enable us to make an informed guess. The heavier, more thickset body type, found in British Shorthairs and Persians, shows the influence of the European wild cat. The foreign and Oriental breeds (such as the Abyssinian and Siamese) retain the lithe body of the African wild cat.

There seems to be no evidence for the claim that some domestic breeds (such as the Angora, Chinese cat and Siamese) have an Asiatic origin and may be descended from Pallas's cat (*Otocolobus manul*) or its close relatives, as the skulls of these cats show no similarity to the Asiatic species.

The development of pedigree breeds

It was not until the middle of the 19th century that the idea of breeding and recording pedigree cats took hold in Britain and Europe. Some breeders started their breeding programmes using ordinary shorthaired 'moggies', selecting them for their body shape and coat

ABOVE: Today there are more than 160 colour varieties of the Persian cat.
TOP LEFT: Blue eyes are the most striking feature of the Siamese cat.
TOP RIGHT: Cats are remarkably agile, and are powerful and accurate jumpers.

colour. From these ancestors, over the years and through selective breeding, today's British and European Shorthair breeds were created.

In America the foundation stock for shorthairs also came from local cats, but these were the descendants of the cats taken over by the early settlers 200 years earlier, and had developed quite distinctive characteristics of their own. These are now reflected in the American Shorthair.

During the early days of cat breeding there were already longhaired domestic cats, but the main development of the pedigree longhair breeds came initially from the Angora cat, which had originated in Turkey, and later from other longhair breeds imported from Persia and Afghanistan. Both the latter types quickly became known as Persians and became popular at the expense of the Angora, which almost disappeared from the breeding scene. By the late 19th century, exports and imports of pedigree cats were starting in earnest, and by the end of that century the Siamese, Russian Blue and Abyssinian had already reached Britain.

During the 20th century the export and import of cats continued. The first Birman arrived in France in 1919 and the ancestor of the modern Burmese entered the United States from Rangoon in 1930.

During the 1950s the Egyptian Mau and Korat reached the United States, and Turkish cats were brought into Britain. The Japanese Bobtail arrived in the United States in 1968, and the 1970s saw the arrival in that country of the Angora and Singapura. Later that century Maine Coons arrived in Australia, where the Spotted Mist was developed, and Ocicats entered New Zealand.

The spread of pedigree cats and the development of new breeds or colour varieties continues throughout the world. There are now dozens of different breeds and hundreds of different colour varieties.

TOP: Cat shows are an ideal venue to familiarize yourself with the characteristics of different breeds. Cats are judged on their condition, their head shape, coat, eye colour and shape, and even their tails.

CATS AND PEOPLE
Cat shows

Cat shows

The first recorded cat show was held at St Giles Fair, Winchester, England, in 1598. During the 19th century, as cats increased in popularity, shows also became more popular. At the earliest events cats were brought along and shown in a variety of containers, or even in their owner's arms. The cat show as we know it today originated from an idea by an Englishman called Harrison Weir, who decided to house and display cats in rows of cages on a bench or table. His first show, called the National Cat Show, was held in 1871 at the

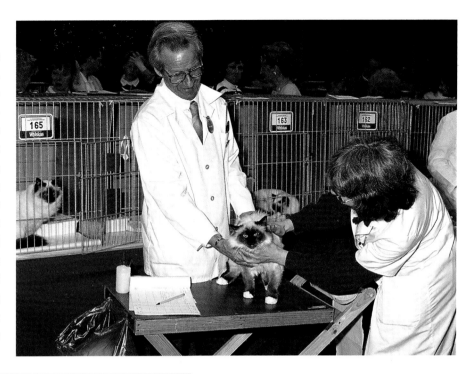

Crystal Palace in London, and at least 160 cats were displayed. This type of display became known as a 'benched' show, a name that is still used today.

Cat shows are now regular events and an integral part of the pedigree cat scene. However, more and more shows are including a section for non-pedigree cats, thereby attracting ordinary cat owners to display their pets and in turn view the pedigree entries. These shows increase the public's awareness of advances in the pedigree world, known as the Cat Fancy.

The Cat Fancy

During the late 19th century, as interest in breeding and showing increased, it became clear that there was a need for some form of control, and for the official recognition and recording of different breeds. In 1887 the National Cat Club, the first organization of its kind, was

ABOVE: This sorrel Abyssinian, a breed that does not like being confined, is clearly an old hand at cat shows.
TOP: Judges examine a British Sealpoint kitten and award points for certain characteristics that are defined in the breed's standard. A kink in the tail would immediately disqualify the cat.

in that country, where the Cat Fanciers' Association (CFA) operates the largest American (and world) registry of cats.

Breed standards

Most countries now have at least one governing body (and some have several) to oversee the recognition and registration of cat breeds and to set the standards for each of them. In each country these standards are contained in special publications that are regularly reviewed and provide guidelines for breeders and judges.

Not all organizations classify their breeds in the same way, and breed standards can vary markedly between one country and another. The biggest variations occur in the United States.

Cats and people

In modern society pet animals are one of the many factors that make up what we call 'quality of life'.

During the last 20 years, numerous studies have confirmed the psychological and medical benefits of pet ownership. These benefits have also become a basis for programmes based on animal-assisted activities (AAA) and animal-assisted therapy (AAT), also referred to as 'pet-facilitated therapy' and 'animal-facilitated therapy'. In these programmes, interactions with animals are used to assist humans with physical or psychological problems.

formed in Britain, with Harrison Weir as its president. It instituted a stud book and set up a system for the official registration of pedigree cats. This organization later amalgamated with another to become the Governing Council of the Cat Fancy (GCCF). In 1983 a further registering body was set up in Britain: the Cat Association of Britain (CAB).

In America the first registering body was the American Cat Association (ACA), established in 1899. Still in existence, it is one of several similar organizations that operate

Animal-assisted activities are informal 'meet and greet' programmes in which any progress on the part of the human recipient(s) is not measured.

Examples are taking pets to visit the elderly residents of nursing homes and hospitals, and the use of pets to help relieve loneliness and isolation in specific groups of humans such as abused children, prisoners, and persons in various forms of therapy.

TOP: Although your cat instinctively hunts birds, it is not impossible for an assertive parrot to learn to dominate a feline member of the household!

CATS AND PEOPLE

Companionship

Animal-assisted therapy is based on a formal programme that sets out to achieve a specific target, and is documented by a professional in the field of health or human services. This person may be a physician, occupational therapist, physical therapist, certified therapeutic recreation specialist, teacher, nurse, social worker, speech therapist or mental-health professional. The animal may be handled by the professional, or by a volunteer under the direction of a professional. The aim of the programme may be improvement in social skills, range of motion, verbal skills or attention span, for example. Each session is documented in the person's record with the progress and activity noted.

For example, an occupational therapist may use the assistance of a cat and its handler in work to increase a person's range of motion in the arm. By making the effort to stroke or hold the cat that person improves his or her mobility. The progress made during each session is documented by the occupational therapist.

In industrialized societies, increasing affluence, a falling birth rate and looser family ties have resulted in pets playing an even more important psychological role. More couples are choosing not to have any children, or to have them later in life after the female partner has established a working career, and for many of these people a pet becomes an important member of the family. But whatever the composition of your human family, owning a cat is likely to provide you with some important benefits.

Companionship

For the majority of cat owners this is by far the most important feature of cat ownership. It is enough simply to have the cat living in the same house, as a partner and a friend in which we can confide. If you talk to your cat as if it were another human being, you are not unusual. Most of us behave in a similar way.

As well as talking to our cats, we instinctively use methods of communication that we would use with other humans. To console them we use standard 'primate gestures' such as stretching out our hands and stroking them, pursing and smacking our lips and 'soft-voicing'.

ABOVE: At seven weeks, kittens such as these are fully weaned and eating solid foods, happy to interact with people and ready to go to a new home.

Comfort, support and relaxation

Comfort may come either from the affection that your cat displays towards you, or from direct physical contact such as when your cat rubs up against you, you stroke it, or it lies on your lap.

Many of us need comforting when we feel sad or depressed, and our cat can certainly help cheer us up. This aspect is particularly relevant to younger family members in times of trouble. If a teenager is going through a particularly difficult period in his or her life, your cat can provide much-needed emotional support.

Your cat will certainly help you relax. It has been conclusively demonstrated that a person in a state of tension shows a slowing of the heart rate and a drop in blood pressure when their pet comes on the scene. Owning a pet is an important stress-management practice for people with high stress levels in their work.

A cat can also provide us with psychological protection. For example, it can give us the emotional security to face or overcome irrational fears, such as a fear of the dark or anxiety at being left alone.

Helping to establish new friendships

There is plenty of evidence to show that people who like animals are more likely to like other people, and are more socially interactive. If you own a cat you are probably good at establishing new human friendships, and are unlikely to allow your cat to become a substitute for, or a distraction from, relationships with other people.

Cats can certainly act as catalysts for contact between humans, and also act as an important link between the young and the old.

Self-fulfilment and self-esteem

We all need to feel good about ourselves. Many of us achieve this through success in our family relationships, work, sport or other recreational activities. Others achieve it through reflected glory by owning or breeding a cat that is an object of prestige. It may be a winner in the show ring, or a rare or unusual breed.

Your cat doesn't have to be a show winner. Every common or garden 'moggy' has a unique character and appearance, and simply by looking at it and knowing that it is yours, you will get a deep sense of satisfaction.

For some of us the mere responsibility of caring for another living creature can result in a sense of self-worth, and by doing it correctly we may be rewarded by the approval of other people.

An aid to leisure activities

Cats are an important part of our leisure experience. They like to play, and they stimulate us to play with them. This helps us to relax and develop a more active zest for life, diverting us from the comparative drudgery of the family chores or work. For many of us merely looking after a cat, such as feeding and grooming it, can become a leisure activity in itself.

ABOVE: An ever-present reminder to relax – a cat's ability to let go and take it easy yet remain alert is the envy of every stressed, overworked cat owner.

CATS AND PEOPLE

Therapeutic value

Benefits to children

The majority of families that own a cat also have children. We might ask ourselves why parents coping with a growing family would want to saddle themselves with another, non-human member, and the answer is not entirely clear. Many of us think that having a pet cat will help teach our children responsibility: that a child who learns to respect and care for a pet is more likely to have a caring attitude towards fellow humans.

There is also an educational value. If our children learn about a cat's body processes and how to cope with its health problems or illness, they may be better prepared for their own experiences later in life. The life cycle of a pet cat averages about 15 years, and may match the period during which our children are growing to maturity. The life of our cat might help to teach them about growing up, learning, old age, suffering and death. Caring for it during that lifetime may teach them some valuable 'parenting' skills.

The presence of a cat in your household can help your children to overcome anxiety, control aggression, develop self-awareness and deal with the problems that occur in life.

Research has shown that when their parents or siblings aren't around, children will often talk to the family cat about the day's successes or failures. It is interesting to note that the children most likely to develop social skills and empathy with other people are those who talk intimately and at length with their pets and their grandparents.

Better housekeeping

It has been shown that families with pets are generally more hygiene conscious than those without.

Therapeutic value

Your cat will probably bring you plenty of other benefits. Statistically you are likely to:

- live longer
- have lower blood pressure
- be in less danger of heart attacks
- suffer less stress and gain more relief of tension
- be emotionally stronger and less likely to become depressed
- have better motivation and be more purposeful
- be less aggressive
- be less self-centred and more supportive of others
- be less judgmental of other people.

Benefits to the elderly

Cats can be of special benefit to elderly people, who often fail to feed themselves properly. Feeding their cat stimulates them to eat, too, and their cat provides them with company while they are doing so.

Elderly people moving into retirement homes would certainly benefit if they could take their pet cat with them, but this is often impractical. For that reason some retirement homes keep one or more cats for the benefit of the residents.

TOP: The Siamese is an extrovert, people-oriented cat and loves human company – it may even be taught to walk on a leash.

YOUR NEW CAT
Choosing your feline companion

Like most people, you may find yourself choosing your own cat, but don't rule out the possibility that your cat will choose you. You will be walking past a pet shop or veterinary clinic and there, peering out at you, will be an adorable little cat face with pleading eyes. You have been thinking about a cat, but not too seriously. Now here is this bundle of fluff, asking you to give it a home. It will be difficult to resist.

Some stray cats have the routine worked out, too. They turn up on a doorstep, reconnoitre the premises, check out the standard of meal service, win over the human inhabitant(s) and then move in.

Although some neighbourhood cats do choose their owners in one of the above ways, most are selected following a planned and carefully considered decision on the part of their owners.

ABOVE: Without a care in the world – cats will leave many of life's responsibilities up to you!
TOP: If you want a pedigree kitten, such as this Birman Sealpoint, make sure you get one from a reputable breeder, and involve the whole family in your selection process.

Factors to consider

- Why do you want a cat? Is it to be a companion, for breeding or to show?
- What other animals do you have already, and will a cat integrate with them?
- Is your property suitable for the type of cat you envisage? A small, high-rise apartment may suit a lethargic domestic longhair, but be inadequate for an active Oriental. A house bordering a busy highway could guarantee a short lifespan.
- Who will look after it? Even if it is a family pet, make sure that one person takes on the responsibility of feeding a proper diet and ensuring that the cat receives the correct vaccinations at the right intervals and is regularly wormed and treated for fleas. Don't rely on the promises of your children.
- How will it integrate with any of the other animals in your household? Will the family terrier terrorize it? Will an incumbent cat consider the new one an interloper and try to drive it away? Is your favourite budgie likely to become a cat's dinner? Will the goldfish in your garden pond continue to lead their current peaceful life?
- Can you afford the cost? Cats may be cheaper to feed than dogs, but they still require health care. This can be costly.
- Is any member of the family an asthmatic? Many asthmatics are allergic to cat fur, so do your homework.

Where to buy your cat

Animal shelters and rescue organizations usually have a selection of cats of varying ages. Many of them employ veterinarians who check the animals' health before they are advertised for a home, and cats from such a source will usually have been vaccinated prior to sale.

Veterinary clinics are another reliable source. Many of them have clients who are looking for a good home for a cat or kittens. These animals will probably have been subjected to health checks, and the veterinary staff will ensure that they receive their proper vaccinations.

ABOVE: Too many kittens are left homeless and have to be destroyed. If you can, get your new cat from an animal welfare centre such as this one.

TOP: Adult cats have less of a chance of finding a new home than kittens do, yet they can be the ideal choice for an elderly person who does not wish to cope with an energetic kitten.

Pet shops commonly have kittens for sale. If you are buying a cat from such a source, do so only if the shop agrees that it is conditional on the cat passing a veterinary health check. If the cat or kitten has not been vaccinated, then ensure that you get this done as soon as possible.

If you decide to get a kitten or cat by answering an advertisement, also only take the animal on condition that it passes a veterinary health check.

Occasionally a kitten or cat may arrive as a (welcome) gift, and in this case the giver should have taken all necessary steps to ensure that the animal is healthy.

Pedigree or non-pedigree?

If you are interested in showing and/or breeding, then a pedigree cat may be the right way to go. If showing your cat rather than breeding interests you, remember that most cat shows have classes for non-pedigree animals, so you don't have to own a pedigree cat in order to be able to show it.

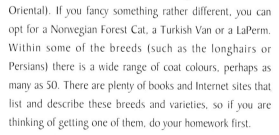

Pedigree cats

Depending on where you live, you may have a choice of 40 or more pedigree breeds, ranging from longhaired cats such as the Chinchilla to sleek shorthairs like the British Shorthair or the Foreign (also called the Oriental). If you fancy something rather different, you can opt for a Norwegian Forest Cat, a Turkish Van or a LaPerm. Within some of the breeds (such as the longhairs or Persians) there is a wide range of coat colours, perhaps as many as 50. There are plenty of books and Internet sites that list and describe these breeds and varieties, so if you are thinking of getting one of them, do your homework first.

Remember, too, to talk to your veterinarian. Vets get to see many of the health and behavioural problems that arise in the local cat population, and have a good idea of any pitfalls. They know which breeders are reputable and which are not, and while they may not be willing to name the latter they can certainly steer you clear of them.

One advantage of choosing a pedigree animal is that you should be able to get a good idea of what the mother, and possibly the father, is like. Reputable breeders are happy for you to visit their breeding cattery and inspect their animals for temperament and health. Avoid a breeder who makes excuses and won't let you view the parents or give you sufficient information.

Just as individual cats vary in their temperament, so do the various breeds. For example, some longhairs (such as Persians) are friendly, comparatively inactive and enjoy nothing more than a cuddle on a warm lap. Others tend to be rather aloof and object to too much handling. The Siamese and Foreign (Oriental) breeds are far more demanding and independent. Don't confuse such perfectly normal 'cat attitude' with poor temperament, though, which is often expressed as aggression. Some pedigree animals have a very poor temperament, caused through selection for their physical appearance with regard to little else, so watch out for it and don't select any animal that comes from such stock.

Cats that have obvious physical defects, such as in-turned eyelids (particularly prevalent in some Persian and exotic varieties) should also be avoided. If a breeder tells you that this condition is normal for the breed, be wary, because while runny eyes or laboured breathing may be acceptable to some breeders, both are a potential health problem. If you are in any doubt, talk to your veterinarian first, or purchase the animal subject to a health check.

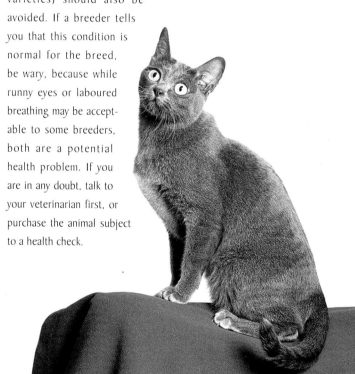

ABOVE: The Korat is a very old breed, native to Thailand (where it is known as the Si-Sawat).
TOP: Cats spend a large part of their day fastidiously grooming themselves, and can reach almost every part of their fur with their tongues. This breed is an Oriental Red Lynx.

YOUR NEW CAT
Kitten or adult?

Non-pedigree cats

The majority of pet cats are of unknown pedigree, and if you choose such an animal the chances are that you will be unable to obtain much information about its ancestry. You may be able to see its mother and get some idea of her character, but that won't necessarily give you an indication as to how her offspring will turn out. Every cat is an individual, and this particularly applies to the non-pedigree or 'domestic' types. What you see is what you get.

Having said that, the vast majority of non-pedigree cats turn out to be ideal household companions. Their ancestors had to be tough, sensible, adaptable and healthy in order to survive, endowing their offspring with what scientists call hybrid vigour: a mixture of genes that gives an individual cat a good chance of surviving and reproducing.

Kitten or adult?

Many non-pedigree kittens are offered for homes from six to eight weeks of age. At this age they should have been properly weaned and socialize well into their new homes. They still need toilet training, and are unlikely to have had any vaccinations.

By contrast, responsible pedigree breeders will not usually allow their kittens to go to a new home until they are at least 12 weeks old. By this time they are house-trained and have received their first course of vaccinations.

A kitten may be far more appealing than an adult, and fulfil your need to nurture a young animal. You will have less idea of what its temperament will eventually be, but the way a kitten is handled and brought up will influence its character, and properly treated the vast majority of kittens turn out be ideal cat companions.

When deciding between kitten and adult, do remember that animal shelters have juvenile and adult cats waiting for, and deserving of, a good home. Some pedigree adults become surplus to breeders' requirements and, after neutering, are available as pets, and it is easier to determine the temperament of an adult cat.

Choosing the sex

If you are buying a pedigree cat and hoping to breed, then you will probably choose a female. A stud cat (male) usually needs to be kept in separate quarters, and because of its scent-marking rarely makes a good family pet. If you do not intend to breed, then sex is not really an issue because as a responsible pet owner you should arrange to have the kitten (or, if necessary, the adult cat) neutered. There is little difference in the behaviour of a neutered male and a neutered female, and both sexes can make an adorable, loving pet.

ABOVE: This Birman breeding cattery is a high-class establishment that provides spacious indoor and outdoor accommodation.
TOP: Play fighting is an important part of the kitten's mental and physical development, where the animals learn to attack others and defend themselves.

Choosing the individual

Look for a healthy kitten or cat whose temperament and personality suits you and your lifestyle. To get an idea of which individual might be suitable, you need to spend some time with it. If you are with an adult cat, sit with it and talk to it, and see how it relates to you. Gauge its reaction to being touched or handled. If you are with a litter of kittens, handle each kitten in turn. A kitten that is unduly shy, or is excessively aggressive towards its littermates, may continue to exhibit those traits in adult life, although that is not always the case. If the kittens' mother is present while you are with them, check her temperament and health as well.

When investigating a cat or kitten for health and temperament, use the following checklist:

- It readily approaches you and does not back away or show aggression.
- It is alert, bright, gentle and playful, and not dull or lethargic.
- It holds its head normally, and walks or runs without limping.
- There is no head shaking, sneezing or coughing.
- The skin appears clean and healthy, without sores, scabs, dirt or flea droppings. The fur is glossy, clean and well groomed, with no areas of hair loss or matting.
- There are no discharges from the eyes, nose or ears. The third eyelid (nictitating membrane) is not partly covering the eyeball.
- The teeth appear clean and free from tartar. The gums are a healthy salmon-pink colour and show no signs of bleeding.
- The belly feels reasonably firm and is not distended. It is neither too hard nor too flabby.
- The anus is clean, and there are no visible signs of diarrhoea or tapeworm segments (which emerge from the anus and look like grains of rice).
- Details of the existing diet are available.

If you don't feel competent to make the above assessments, take somebody along with you who is. If even one member of a litter of kittens appears unhealthy, you would be wise to choose from another source.

Once you have decided on an individual, make sure that it has been sexed correctly.

Finally, ask for a 10–14 day approval period during which you can obtain an independent health check from your veterinarian. Any infections that are incubating will show up during this time.

ABOVE: Kittens that have had no contact with humans may be distrustful and difficult to socialize.
ABOVE RIGHT: This cat's raised tail indicates that it is confident and willing to interact.
TOP: Take your kitten to the vet for a health check, as a runny nose and eyes could indicate a problem.

YOUR NEW CAT
One cat or two?

If you have purchased a pedigree cat or kitten, make sure that you receive the correct registration papers. If one of the conditions of purchase is that the cat or kitten must be neutered (de-sexed), then it is normal practice for the breeder to withhold such papers until after you provide proof that the operation has been performed.

One or two?

Many people think in terms of one kitten or cat, but do consider the option of taking two together. Cats certainly enjoy solitude on occasions, but they are also communal animals and two individuals will often prove good company for each other while their human owners are away from home.

Integrating a new cat with existing pets

Before you bring a second (or even a third) cat into your household, you should make sure you are familiar with cat behaviour, both territorial and aggressive, and the basic principles of cat training (see pp61–71). Some cats will accept a newcomer, especially a kitten (which may be perceived as less of a threat), but others will not.

You will help this integration by gradually introducing the newcomer to your cat(s), keeping it separated in one room until it has gained confidence (especially if it is a kitten) and your existing cats have got used to it. The newcomer is probably unsure and may be frightened, and it is moving

TOP: Cats and kittens may not usually like water, but they certainly enjoy stalking its inhabitants.

into new and unfamiliar territory which is already occupied (and maybe defended) by the current feline inhabitants. Feed them separately to reduce competition for food, and make a fuss of your existing cats so that they do not feel neglected because of the new incumbent.

If you already own a dog, the introductory process is similar to that described above – gradual and non-threatening. Once again, the dog may immediately accept the cat. Sometimes a bitch will accept a kitten and relate to it rather like the way she would to one of her own pups, even to the extent of offering it some protection. Make sure that you give the incumbent dog as much fuss and attention as usual (or even more), and praise and reward it for good behaviour.

Introducing a new cat to pet birds can pose a problem. If it is a kitten purchased from a breeder, it may never have experienced the sight or stimulation that a bird presents. Although its basic hunting or playing instinct may cause it to react, it may be quite easy to train your cat to ignore the bird or even to accept it as a companion. If, however, it is a kitten from a domestic cat that has had the opportunity to introduce its kittens to bird prey, or is an adult cat that has already learned to catch birds, then you have a more difficult task on your hands. If you find that you do have such a problem, talk to your veterinarian.

ABOVE: Dogs, especially bitches, may willingly accept and mother kittens in the household.
TOP LEFT AND RIGHT: Goldfish and small mammals that are kept as pets need to be protected from the family cat – make sure their cages are secure, and avoid keeping a goldfish in an open bowl.

YOUR NEW CAT

Introducing a cat to children

Goldfish are yet another pet that can be threatened by an incoming cat. Those kept indoors in an aquarium tank with a glass lid and artificial lighting should be safe, but any that are exposed to an inquisitive cat may stimulate an unwanted reaction. Goldfish in an outdoor pond are also susceptible to a cat's attentions, and you may need to train your cat to leave them alone. Protective measures include physical barriers such as netting, and the installation of plenty of water plants, such as water lilies – the fish can hide under the leaves.

Introducing a cat to children

If you have a baby, you are more likely to want to protect it from the cat. Make sure that the cat cannot climb into a young baby's cot, because there is always the danger that the cat will jump down onto the baby's face and scratch it, or curl up close to the baby's face and obstruct its breathing.

Toddlers can cause a new cat some problems, because they tend to want to hold the animal – usually in an extremely uncomfortable, if not painful, position. You need to train your child just as much as your cat to ensure that they both get the most enjoyment from each other.

The same applies to older children, especially if this is the first pet they have experienced. They need to understand how the newcomer feels, and the importance of keeping it free from stress and allowing it some time out on its own. Children have similar needs, and it shouldn't be too difficult for them to learn to treat the new cat as an individual rather than an object or toy.

ABOVE: When choosing a new kitten, try to view the whole litter at home. Although the smallest kitten may look cute, it is more likely to be weak and sickly. The boldest, biggest kitten is a better choice.

CARING FOR YOUR CAT
A new cat, a new home

Ideally every cat should live in the home of a caring and informed owner, but many cats are left to fend for themselves or are treated as dispensable items that can be left behind when the owner moves house. If you regard owning a cat as a privilege, not a right, then you will help to create the happy and satisfying mutual relationship that so many cat owners have experienced.

Remember that your cat's temperament, enjoyment of life, health and welfare are influenced by the way you and your family treat it. Properly cared for, your cat can bring you happiness and provide you with affection and loyalty. Many cats eagerly await their owner's return home at the end of the day. Some will accompany their owner on an evening walk. Some even show their appreciation by bringing home prey and presenting it to their favourite person. Yes, believe it or not, cats can indeed be loyal to their owner.

ABOVE: When you hold your cat, always keep one hand under its hindquarters to support its weight.
TOP: A longhaired cat may be soft and cuddly, but it will need daily grooming – up to 20 minutes at a time – to stop its coat from matting.

A cat bed and bedding

It is possible to purchase a wide variety of manufactured plastic or wooden cat beds, wicker baskets or bean bags, together with washable wool or synthetic rugs.

If you want something cheap but effective, an old cardboard box with one side partly cut away to make an entrance will prove perfectly adequate. Line the box with newspaper and place a piece of washable blanket on top of the paper. Remember, though, that a cardboard box cannot be properly cleaned and will need to be replaced from time to time.

Place the bed in a quiet, draught-free corner away from family traffic so that the cat can have some privacy when it so chooses. If you have a spare room, lend this to your new cat for a week or two until it has grown used to your household and its activities.

A litter tray and cat litter

Most importantly, a cat litter tray should be easy to clean. Some trays have disposable liners, although newspapers will serve the same purpose, but they will need to be changed at frequent intervals or they will smell. It is best to use a commercial cat litter containing absorbent clay material or Fuller's earth, substances that absorb the odour of urine and faeces. You can find cat litter made of bark, but this is not as effective.

Place the litter tray in a quiet corner, well away from the cat's bed and its food: cats will not perform their toilet close to where they eat.

Feeding utensils

Food and water dishes must be either washable or disposable. You can use disposable plastic dishes or one of the many types of plastic or pottery bowl. Plastic dispensers for dry food and drinking water are ideal once a kitten or cat has learned to use them. Whichever type of utensil you choose, it is important to clean or change them regularly.

Grooming gear

Basic equipment comprises a double-sided body brush as well as a coarse and fine (flea) comb. If you have a long-haired cat you may require additional equipment, such as

TOP: Even though your cat will enjoy sleeping on your bed or any cushion or chair, it is important that it does have a bed of its own, to provide a sense of security especially when it is feeling ill or insecure.

toys are rather frustrating, although balls that roll along provide them with the opportunity to chase, and some cats seem to enjoy the challenge of trying to grip them.

Toys that stimulate through sight or sound are also useful – look for the brightly coloured 'wands' that are often used by breeders to get their cat looking animated in front of a judge, and by cat photographers to capture that appealing portrait.

Whatever type of toy you buy, make sure it doesn't have any small metal or plastic pieces that could be chewed off and swallowed.

You certainly don't have to spend much money on toys. Small balls of rolled-up newspaper, or pieces of cloth attached to a length of string, will serve the purpose, although many cats soon tire of a toy that they deem to be too artificial. A lump of fur or a feather is far more stimulating, although in the paws of an agile cat the latter won't last very long.

a pair of blunt-ended scissors for cutting away sections of a matted coat. If you're an eager groomer, keep a chamois leather cloth to polish your shorthair's coat.

Collar

If you wish your cat to wear a collar with an identity disc or a bell (to warn garden birds of the cat's presence), attach it once your cat has settled into its new home.

The collar must have an elastic section to stop the cat from getting caught by the collar and choked.

Toys

Visit any pet shop or supermarket and you will see a vast array of cat toys, such as fluffy or plastic balls.

Cats generally enjoy soft toys that they can grip with their claws; hard plastic

A scratching post

Claw-marking – when cats use their claws to scratch objects – is not confined to domestic cats: lions, tigers, leopards and many other types of cat go through a similar routine.

Claw-marking has two functions. One is territorial, and the claw marks could be described as 'cat graffiti'. The scratches are a visual signal that another cat owns or visits this particular piece of territory. A certain amount of scent, left on the object from the glands on the footpads, reinforces the signal and identifies the individual that left it.

ABOVE: Take care that your cat's toys have no small pieces that could be chewed or bitten off, then swallowed.
TOP: A grooming tool for every coat type: rubber brushes, wire bristle brushes and fine- and wide-toothed combs.

CARING FOR YOUR CAT

Claw-marking

The second function of claw-marking is cosmetic, because the scratching action helps to remove dead layers of the protein keratin from the surface of the claws, thereby keeping them sharp and in good working order.

The innate need to claw-mark can get some cats into trouble, especially if they are not provided with a suitable object, such as a scratching post, on which to carry out this important function.

You can buy a ready-made scratching post, although it is simple to make your own from a soft wood such as pine, and cover it with carpet offcuts, textured cloth or bark. The post must have a heavy base to prevent it from falling over as a result of the cat's vigorous scratching.

Even when a post is available, the best lounge suite may prove more attractive – you will need patience and your cat will need training (see pp69–70). Initially the scratching post should be placed close to the furniture, and the latter protected to discourage the cat and prevent further damage. Because scent is left on the furniture during the scratching action, this needs to be masked by the use of a deodorizing agent or repellent spray. The scratching post can be gradually moved away from the furniture to a mutually acceptable place.

ABOVE: Teach your cats to use a scratching post as soon as possible – it will be difficult to persuade them to leave the furniture alone after they've had this freedom for a few years.
TOP: Lions, like domestic cats, mark their territory with scratch marks on trees and other objects.

A cat door

If your cat is allowed access to the yard or garden, a cat door is highly recommended. One possible disadvantage, though, is that neighbouring cats might also learn to use it, and pay unexpected and annoying visits and eat your cat's food.

To prevent this problem, it is possible to buy a sophisticated type of cat door where your cat wears a collar with its own electronic 'pass' system.

A carrying basket

A carrying basket is not essential, but very useful. There are various types, from folding cardboard designs (which cannot be cleaned and have a limited life) to permanent models with plastic trays with wire tops and lids, or fully moulded bodies with air holes and front-opening doors.

Arriving home with your new cat

If you have brought home a kitten, in most cases it will just have been taken away from its mother or littermates, and that companionship will need to be replaced. One or more members of the family should act as a surrogate companion for as much time as possible. That means lots of cuddles by responsible people, but only when the kitten wants them. Handle it gently, and restrict the amount of handling that it gets, especially by children. If you have brought home two kittens, then they will be company for each other.

Introduce new experiences gradually, and avoid stressing your kitten or cat with loud or sudden noises. It will usually take several days for a new kitten or cat to adapt to the sights and sounds of your household environment.

ABOVE: A good cat carrying basket must be strong and secure, well ventilated and easy to carry and clean.

TOP: Make sure your cat flap is about 6 cm (2.3 in) above the floor indoors, so the cat is able to simply step through it.

CARING FOR YOUR CAT
Arriving home with your new cat

Food and water

For the first few days at home you should feed your new cat the same diet as fed by its previous owner. After that, if you wish, you can gradually introduce a new diet over a period of four days, replacing about a quarter of the old diet with the same amount of the new one each day.

Make sure that clean, fresh water is always available. This is especially important if your cat is being fed a high proportion of dry food.

Safety

When preparing for a kitten's arrival, think about safety in the home, just as you would for a young child.

- Lock away all household chemicals or poisons. Although kittens are far less inquisitive than puppies and much less likely to ingest poisonous substances, it is best to be on the safe side.
- Make sure that there are no frayed or bare electrical wires that a kitten could chew.
- Be aware of some of the risks associated with some common garden plants.
- Remember that sparks from a fire or cigarette ash can burn eyes or skin.
- Make sure anyone using mowers, bicycles, skateboards, roller blades or similar articles is extra vigilant.
- Check where the kitten is before moving a vehicle.
- Make sure the kitten cannot get through the fencing around a swimming pool.

Infections you could catch from your cat

Some cat infections, such as the roundworm *Toxocara cati*, ringworm infection and *toxoplasmosis*, can be passed on to humans. And if your cat scratches you, you could get cat-scratch fever. For more details, see pp84–95. Explain the risks to all family members, and insist on basic hygiene, such as the regular washing of hands and immediate cleansing of any wounds. A doctor should see any deep wounds inflicted by your cat.

ABOVE: Territory is very important to cats, and windows serve as a good vantage point for a cat to survey its 'property'.

House rules

As a member of the family, your kitten or cat must learn its place. Because cats are more independent and solitary than dogs, they don't easily fit into a 'pack' order. Nevertheless, they will learn to obey commands and accept that human members of the family are to be treated as dominant.

Teach your cat or kitten basic house rules, for example, that it may not jump onto the kitchen counter or the dining table, that it may not beg at the table, and that it cannot always have its own way.

Toilet training

Cats are naturally clean animals, and if you have provided a litter tray sited in the correct location away from food, an adult cat will quickly learn to use it. Kittens need to be taught to use a litter tray, though, and you can do this by placing them into it when they show signs of wanting to perform. Even better is to anticipate when they might do so. Toiletting occurs most frequently soon after waking up and after meals, and if you place a kitten into a litter tray at these times it will soon learn to perform on cue.

Territory

Your new cat or kitten needs to learn the extent of your territory, and come to terms with neighbouring cats and which cat owns what.

Start by keeping your cat shut in your home (even in one room, if necessary) until it feels secure. Then, if you are able, allow it to venture outside. Try to allow it to establish its own territory and make its own peace or friendship with neighbouring cats. Some of these may already regard your garden or yard as their own, and strongly object to the presence of your newcomer. You may be able to help your new cat to establish and defend a new territory by discouraging intruders, but in many cases a better and more permanent solution is to let the cats sort out the problem among themselves. There may be lots of noise for a while, but hopefully there will be few, if any, battle scars.

Vaccinations

The usual age for a kitten to commence its course of vaccinations against common viral diseases such as cat 'flu and snuffles is between nine and 12 weeks, although under special circumstances the first vaccination can be given at six weeks.

If you have purchased a pedigree kitten it may already have completed its initial vaccination programme, and an adult cat obtained from a rescue centre should already have been vaccinated. A separate vaccine is used to cover feline leukaemia. Talk to the staff at your veterinary clinic to find out what vaccines are needed in your area.

ABOVE AND TOP: Cats mark their territory by claw-marking furnishings or wallpaper, or by rubbing the head and face against an object, depositing scent from the sebaceous glands.

CARING FOR YOUR CAT
De-sexing

De-sexing

Neutered cats of either sex make the best pets. There are already thousands of unwanted kittens needing good homes, so unless you are breeding from pedigree cats, please don't add to their number.

Ask your veterinary clinic about the best age to have your kitten neutered. Most veterinarians recommend that females be spayed at 24–30 weeks old, before they commence their first oestrus (that is, 'come into season'). The average age for oestrus is about six months, but some individuals of 'precocious' breeds such as the Siamese may start 'calling' as early as four and a half months. Some cats don't have their first oestrus until they are nine months old, or even older.

Males can be castrated from 16 weeks, but many vets prefer to leave them until they are about six months old. They believe that this allows more time for the development of the urethra, which leads from the bladder through the penis, and thereby reduces the likelihood of the blockage that causes the condition known as feline urological syndrome (FUS). Males should certainly be neutered by nine months, before they have fully developed their male characteristics, learned to roam and started fighting.

Worming

Various types of worm can infect kittens and cats. When you get your kitten or cat, it may already have been wormed. If not, it may need treatment. Either way, as soon as you get your cat, talk to your vet about what worms are prevalent in your area, what treatments are recommended and how often you should use them. For further information, see pp93–5.

Flea control

Fleas are a common problem and your kitten or cat should have been treated and be free of fleas before you obtain it. However, it will need further, regular treatment against them. For further information, see p92.

Exercise

Kittens and cats usually exercise themselves, and you will contribute when you become involved in play sessions. If your cat becomes lazy, you may need to encourage it to play.

Training

Basic training for cats involves toilet training and obeying house rules. It is also useful to train your cat to travel in a cage, and to get used to being parted from you. Then, if you need to take it to the vet or to a boarding cattery, it should be far less upset or stressed. This type of training is especially useful if your cat is particularly shy or nervous.

ABOVE: Although cats are essentially solitary creatures, they do enjoy the company of other cats, especially if they have been brought up together.

TOP: It is not merely an 'old wives' tale' – cats have a superb sense of balance and usually do land on their feet.

First, you must get it used to being put into a carrying cage. Initially, leave it in the cage in one of your rooms for a few minutes, then take it out and reward it with food or a cuddle. Extend the period of caging until it is content to remain in the cage for up to an hour.

Next, put it in the cage, carry it outside and place it in a vehicle parked in a quiet spot. Make sure there is adequate ventilation. Leave the cat there for a short time, then bring it back inside. Release it and reward it.

Gradually extend the time you leave it in the vehicle. Finally, if you have a willing relative, friend or neighbour, leave it in the cage with them for short periods. This gets your cat used to being separated from you, and it learns that you will be coming back.

Pet cats can be taught (or sometimes teach themselves) simple tricks such as pulling on a handle to open a door. If you watch advertisements for pet food, videos or movies, you will be aware that cats can be trained to perform, but this is a specialist area in which you are not likely to become involved. If you do wish to do so, you need to get expert advice.

For more information on training and behaviour, see pp61–71.

Teething

Between the ages of 14 weeks and six months your kitten's temporary (milk) teeth will gradually be shed and replaced by permanent teeth. Shedding normally starts with the incisors, followed by the premolar, molar and canine teeth.

Your cat will usually get through this process without you even noticing, and won't need any special help with its diet. If it does seem to have a problem when eating, though, provide a more moist diet. If this doesn't solve the problem, talk to your vet.

Dental hygiene

For cats, as with humans, a proper diet that includes plenty of dry or chewable food will help to keep teeth clean and gums healthy. Nevertheless, tartar may begin to accumulate on the teeth, especially as a cat gets older, and they should be checked regularly.

Grooming

Cats groom themselves to keep their fur clean, but also to regulate body temperature. Because of their thick fur, cats' sweat glands are not as effective as those in a human. In hot weather or after strenuous activity, a cat cannot lose enough heat by this method, so it compensates by licking saliva onto its fur: the saliva evaporates and helps to keep the cat cool. This explains why a cat grooms itself more after spending time in the sun and after exercise such as playing or hunting.

Licking the fur also stimulates sebaceous glands in the skin. These secrete an oily fluid that helps to keep the cat's fur waterproof – the fluid also contains a small amount of vitamin D, which the cat then ingests.

ABOVE: Patience and practice might persuade your Siamese, Burmese or Russian Blue to walk on a harness.
RIGHT: Grooming keeps the cat's coat glossy and clean, and it also stimulates blood circulation.

CARING FOR YOUR CAT

Grooming

Most cats need very little grooming help from their owners. Some are lazy, though, and don't groom themselves enough. If you wish, you can stimulate such a cat to groom by spreading a little butter onto its fur.

Some cats can't groom themselves properly because of their long hair or because of old age. You will need to groom such a cat regularly.

Brushing and combing

Get your cat used to being handled and groomed. Establish a daily routine in which the cat is gently placed onto a non-slip surface (a piece of old carpet or something similar) on a table, and rolled over to have its mouth, teeth, eyes, ears, abdomen and paws examined.

Although it may not need grooming, do it anyway. It will help to train your cat and you will more quickly detect fleas or flea dirt, and any hair or skin problems. Try to make each session pleasant for the cat, and praise and reward it for good behaviour. Your basic grooming equipment should include a cat brush, comb, grooming glove (mitt), sponge, cotton balls, cat towel, blunt-ended surgical scissors and (if you wish) nail clippers.

There are various types of cat comb. Some have wide teeth, and can be used on long, fine coats. Some have teeth of varying length, and others have only fine teeth (flea combs). There is also a special type of comb with wire projections for use on a thick undercoat to remove tangled hair, as well as a 'slicker' brush for use on the tail, especially before a show.

When grooming a longhaired cat, pay special attention to the feathering on the legs and the tail. Matted fur may occur in areas that the cat cannot easily reach to groom, such as on the inside of the elbows and along the abdomen close to the thighs. Also check the paws, nails and paw pads. In longhaired cats, hair may grow beyond the level of the pads. If so, trim it away using a pair of blunt-ended, curved surgical scissors. Also check under the tail, wipe away any debris and cut away any excess hair.

Use a damp cotton wool ball to wipe away 'sleep' from your cat's eyes.

ABOVE: When grooming your cat, it is important to place it on a non-slip surface, such as this piece of carpet.

Cat groomers and grooming parlours

Many pedigree cat breeds, especially those with long coats, require considerable grooming. If you know what to do and you have the time, you can do this yourself. If not, you can get an expert to do it for you.

If you would like to learn to groom your cat, ask at your nearest grooming parlour or vet for details of grooming schools and courses.

Bathing

If you carry out regular grooming, you should need to bath your cat only if it becomes particularly dirty or smelly, or prior to a cat show.

Always give your cat a thorough brush-out before bathing it. Use lukewarm water, which is more comfortable, and a proper cat shampoo. Don't let shampoo get into any body opening. Rinse thoroughly, paying special attention to the areas between the forelegs and hind legs.

A cat can easily become chilled when wet, so make sure you dry it properly afterwards, using its own special towel. If you prefer to use a hair drier, run your fingers through the cat's hair as it is being dried to make sure that the air stream is not too hot.

Nails

A cat's nails, like those of humans, are continually growing. The action of claw-marking is usually enough to keep them worn down and in good shape, but in some cases the nails will need cutting with nail clippers. You may be able to do this yourself, or you may prefer to ask someone at your veterinary clinic or a cat groomer.

ABOVE: When a cat scratches a tree, an old claw sheath may come off and be left in the bark.
TOP: Some cats do tolerate being bathed, especially if introduced to the procedure gently and while they are still kittens. Take care to dry your cat thoroughly afterwards, as cats easily become chilled.

CARING FOR YOUR CAT
Travelling with your cat

Travelling with your cat

Whenever you take your cat away from home, make sure it is wearing a suitable elasticated collar that bears a tag with your name and telephone number, or an electronic identi-chip (a relatively cheap, effective and increasingly common hi-tech method of keeping track of your pet).

In your car

If possible, train your cat from kittenhood so that it gets used to travelling. Early training will help eliminate any fear or agitation, and reduce the likelihood of motion sickness.

Don't let your cat have complete freedom within a vehicle. It can distract the driver and lead to an accident, and the cat is at risk of injury if an accident occurs. For its own safety the cat should be in a plastic or metal travelling cage, which should be fastened by a seat belt as a precaution against an accident.

If your cat is likely to travel in your car quite often, train it to do so as soon as you can. Your cat's travel cage eventually becomes an extension of its home territory, and it will feel comfortable inside and readily occupy it. Travel cages are commonly used by people who show their cats, and are an ideal way of providing a safe, secure, 'personal space' for your feline companion. Once your cat is accustomed to it, the cage can accompany it wherever it goes, providing it with a 'home from home'.

If you have to leave your cat in your car, make sure that the vehicle is parked in the shade and has adequate ventilation. In the sun, the temperature inside a closed

ABOVE AND TOP: Many cats learn to travel well, if started early as kittens. If your journey is to last for longer than an hour, schedule regular breaks to allow your cat to eat, drink and use its litter tray.

car can quickly exceed 40°C (104°F), and heatstroke will set in very rapidly. Do not assume that a car parked in the shade will remain so: as the sun moves around, a shady area may become fully exposed. Special screens can be fixed to open windows to provide car security and enough ventilation.

On holiday

Make sure that your cat's vaccinations are up to date, as the incidence of infectious disease in your holiday area may be greater than in your home neighbourhood. Different sorts of external parasite, such as ticks, may also be present, so thoroughly groom your cat every day and check its skin for their presence.

If you are staying in one place, make a note of the nearest veterinary clinic.

In a bus, train or plane

If you are taking some form of commercial transport, your cat may be required to travel separately in a cage. This can be a frightening experience for a cat that has not been cage-trained, so anticipate the event and train your cat to feel safe in its own crate, with its own familiar toys and bedding.

Don't feed your cat within six hours of the start of the journey. If you think it may suffer from motion sickness, ask your vet for advice.

Travelling between countries

Travel between countries usually involves travel documentation for your cat as well as yourself. Regulations vary, so make sure you know what they are for the particular country you are heading for. You will probably need a veterinary certificate stating that your cat is fit to travel and is free from any infectious or contagious disease. You will also require an up-to-date rabies vaccination certificate. Many countries require these documents to be in their own language.

There are many countries where rabies does not exist. These include the United Kingdom and certain European countries. Some islands, such as Hawaii, Australia and New Zealand, are also rabies-free. Some countries require a cat to be quarantined upon entry, others will allow entry providing that certain conditions, such as microchip identification and blood testing, are met (see the PETS Travel Scheme, below).

When making plans for such travel, contact the consulate or embassy of the country concerned. Information may also be available on the Internet.

If travelling abroad requires you and your cat to be separated, carry out the training procedures recommended for commercial travel on a bus, train or plane (see above).

The PETS Travel Scheme

A pilot scheme for the issue of pet passports was introduced in Britain late in February 2000. Under this scheme, cats and dogs are allowed to travel from the British Isles to specified countries in western Europe and return home without having to endure six months quarantine upon their return. Owners must use designated carriers and ports of entry, and cannot import a pet under the PETS Travel Scheme from a private boat or plane.

Cats and dogs resident in certain European countries that are taking part in the scheme are also allowed to enter the British Isles. At the time of writing these countries were Andorra, Australia (Guide dogs and Hearing dogs only), Austria, Belgium, Denmark, Finland, France, Germany, Greece, Iceland, Italy, Liechtenstein, Luxembourg, Monaco, Netherlands, New Zealand (Guide dogs and Hearing dogs only), Norway, Portugal, San Marino, Spain, Sweden and Switzerland.

TOP: Travelling overseas with your cat requires an enormous amount of organization, not least having to make sure that you have the correct legally specified container for your animal.

CARING FOR YOUR CAT
Travelling between countries

Cats resident in the British Isles

To qualify for a passport, a cat resident in the British Isles must have an identification microchip inserted under the skin. Once the cat is at least three months old an approved veterinarian must then vaccinate it against rabies. Some time after the last vaccine injection (30 days is the ideal period), a blood sample is taken by the veterinarian and sent to one of the government-approved laboratories. Once the sample passes the test, a health certificate or passport is issued and stamped by the veterinarian.

Between 24 and 48 hours before returning to Britain, the cat must be treated against a particular tapeworm and ticks, and a veterinarian approved by the relevant government must issue a health certificate.

Once a cat has been vaccinated against rabies, booster vaccinations are required every year.

Cats resident in designated European countries

Animals resident in specified countries in Europe can also qualify to enter Britain if their owners follow the same rules. However, their owners must wait for six months from the time a successful blood test sample was taken.

Cats resident in the United States and Canada

Because rabies is endemic in North America, the original PETS Travel Scheme did not apply to cats entering from that region. At the time of writing these cats must still endure a six months quarantine period in Britain, although the situation will be reviewed once the success of the initial scheme has been assessed.

Cats resident in rabies-free islands

If it proves successful, the scheme will be gradually extended. Providing that the relevant veterinary authorities and the airline carriers agree, cats and dogs may be allowed to travel between Britain and designated islands that are rabies-free.

Boarding catteries

Most cats adjust very quickly to going to a boarding cattery. The standard of catteries varies and usually (but not always) you get what you pay for, so the more expensive the fees the better the quality of comfort and service you should expect. A reputable cattery will allow you to inspect its facilities beforehand. If you do so, observe how the resident cats are behaving and talk to the staff about feeding, grooming and exercise routines.

The staff at your local veterinary clinic may be able to recommend a suitable cattery, and will advise on vaccination procedures. Reputable catteries require their boarders to have up-to-date vaccination certificates against common infectious diseases.

Cat-sitters

If you don't like the idea of boarding your cat, you may wish to employ a cat-sitter or house minder to provide live-in care of your home and your cat while you're away. Your vet should be able to give you contact details.

ABOVE: Before you book your cat into a boarding cattery, visit the premises to check whether they are clean and spacious.

NUTRITION
A balanced diet

Like all animals, the domestic cat needs a diet that is properly balanced and contains all the essential nutrients in the correct quantities. These nutrients are water, protein, fat, carbohydrate, minerals and vitamins.

The wild members of the cat family (*Felidae*), such as the lion, tiger, cheetah and European wild cat, are carnivores. Between them they hunt and kill a wide variety of other animals, ranging in size from small lizards and birds to large antelopes. They don't just eat the meat or muscle, but consume all, or almost all, of their prey, including the skin, hair or feathers, and the internal organs such as liver, kidney and intestine. Their diet therefore contains a substantial amount of animal protein, and supplies them with all the other essential nutrients that they require.

To remain healthy, domestic cats also require a diet containing animal protein. This is because they need a particular amino acid (one of the building blocks of protein) called taurine, which helps to prevent heart and eye disease. Taurine is plentiful in animal protein but only present in small amounts in plant protein.

While dogs are able to manufacture the amino acid taurine within their body, cats can only manufacture a little, and it is not enough to meet their needs, and plant protein cannot supply them with enough to make up the shortfall. Therefore although a pet dog could remain healthy if fed a properly balanced vegetarian diet, a cat could not.

For this reason cats are known as obligatory carnivores; they must eat some animal protein in order to survive.

Water

This is the most important element in a cat's diet. Whereas an animal can survive after losing half of its protein and storage fat, even a 10 per cent loss of total body water causes serious illness, and a 15 per cent loss causes death.

Animals obtain water in three ways. They drink it, eat food that contains it, and their body manufactures some water during the chemical processes involved when converting proteins, fats and carbohydrates into energy.

The daily amount of water required by a cat is roughly the same amount (in millilitres) as its energy requirement (in kilocalories) (see the table on p47). A sedentary cat needs a daily intake of about 65–70 ml (roughly four tablespoons) water for each kilogram (2.2 lb) of body weight, while an active cat needs about 85 ml (roughly six tablespoons).

TOP: Unlike wild lions, which take a variety of prey and eat most parts of it therefore ensuring a balanced diet, the domestic cat relies heavily on food provided by humans.

ABOVE: This leopard will maintain healthy teeth and gums by chewing on flesh and bone. Domestic cats also need to chew to maintain dental health, which is an important consideration when planning your cat's diet.

Protein

Protein occurs in animals (animal protein) and in plants (plant protein). There are many different types of protein, each of which contains a particular combination of amino acids, the substances that provide the materials needed for the growth and repair of all body tissues.

Proteins vary in their digestibility. The most digestible are those contained in foods derived from animal sources, such as meat, eggs and cheese. The least digestible are those contained in foods derived from plants, such as grains and vegetables. Most domestic cats consume a diet containing a significant amount of animal protein. They do eat some plant material, either in the stomach and intestines of prey that they catch, or by voluntarily eating specific plants such as grass, but plant protein is a comparatively unimportant part of the domestic cat's diet.

When a cat eats grass it is probably doing so to consume fibre and as an aid to digestion. Quite often a cat will vomit soon afterwards, bringing up a bolus of grass mixed with mucus, so eating grass may be a useful method of getting rid of excess mucus from the cat's stomach.

Fats

Fats and oils contain substances called fatty acids, some of which play an important role in helping to maintain certain internal body functions and a healthy skin. They also act as carriers for the fat-soluble vitamins (A, D, E and K). Fats are a concentrated form of energy (for a given weight, fat provides more than twice as many kilocalories as carbohydrate or protein).

If the diet that a cat consumes contains more energy than the cat needs, the excess is converted into fat that is stored in various parts of the body, such as under the skin and around the intestines. This stored fat acts as a fuel store that can be drawn upon in times of need.

Carbohydrates

These occur in plants and include sugars, starch and cellulose. There are various types of sugars, among which are sucrose and glucose, two of the simplest sugars and therefore more easily digested. Cow's milk contains the milk sugar lactose, but many adult cats are unable to properly digest lactose – for this reason specially formulated lactose-reduced or lactose-free cat milk is available from pet food stores and supermarkets. For cats, one of the most useful sources of dietary carbohydrate is rice.

Minerals

Like other animals, the cat needs to consume many different minerals to ensure that its body processes function normally. Some are required in comparatively large amounts, while others, known as trace elements, are only required in very small quantities. Two of the most important are calcium and phosphorus, involved in the formation and growth of bones and teeth. Minerals play an important role in the growth and repair of body tissues such as muscles, ligaments, skin and hair, the formation of red and white blood cells, and in various digestive processes.

Vitamins

Certain vitamins are essential for the proper working of body processes. Four of them, vitamins A, D, E and K, are soluble in fat, so fats and oils provide a good dietary source.

NUTRITION
Energy requirements

Vitamins A and D play a particularly important role in bone growth. Vitamin E plays an important role in normal muscle function, vision and reproductive processes. Vitamins of the B-group, and vitamin C, are soluble in water. The B-group vitamins have a variety of functions associated with the metabolism of amino acids, fats or carbohydrates. Vitamin C is involved in wound healing, preventing haemorrhages from small blood vessels (capillaries) and maintaining healthy skin (preventing scurvy). Cats, like dogs, can manufacture this vitamin within their body, and unlike humans, they don't need a source of vitamin C in their diet.

Fibre

Derived from plant materials (often eaten along with prey), fibre does not provide a cat with any nutrients but it does play a very important role in digestion. It acts as a bulking agent, absorbs any toxic by-products of the digestive processes, and increases the rate of passage of food through the gut.

Energy

Energy is measured in kilocalories. It is not classified as a nutrient, but is the 'fuel' that a cat derives from the protein, fat and carbohydrate that it eats. Cats require sufficient kilocalories to fulfil their basic energy needs, and this amount varies according to their size and circumstances. Adult cats do not have such a great range of sizes and weights as adult dogs. There are variations between breeds and individuals, but most adult domestic cats weigh between 2.5 kg (5.5 lb) and 5.5 kg (10 lb). Sedentary house cats need less energy than active cats that spend a lot of time out of doors. Energy use also varies according to environmental temperature, being greater in very cold temperatures or in hot tropical climates. Comparatively more energy is required per kilogram of body weight by a queen in late pregnancy and during lactation, by a kitten when growing, and by any cat during illness or stress.

Given the opportunity (for example, self-feeding on dry pet food and/or freedom to hunt), some cats will feed little and often, exercise themselves and remain within a satisfactory weight range. Others will eat everything on offer, and if their owner does not monitor their energy intake they will quickly become obese.

A guide to energy requirements for cats

ACTIVITY	APPROXIMATE DAILY ENERGY NEED
	Kilocalories per kg (2.2 lb)
Sedentary	65–70
Active	85
Gestation (last 3 weeks)	90–100
Lactation	140–170 (depending on the number of kittens in the litter)
Growth (weaning to 6 months)	130
Growth (6–12 months)	100

It is not unusual for domestic cats to put on weight in autumn and winter and lose it again in the summer. This probably reflects the situation in the wild, where many animals lay down a store of fat prior to winter when food becomes scarcer. When cats put on excessive weight and keep it on, it is usually the result of overeating or lack of exercise, or a combination of both. In this case their energy intake should be strictly monitored, because overweight cats, like overweight humans, are more likely to have health problems.

ABOVE: Most cats are disciplined eaters and will eat little, often. Some cats, however, will eat all the food in their bowls, no matter how much it might be. Owners of such cats need to monitor their pets' energy intake, or the cat may soon become obese.

Commercial or home-cooked food?

Fast foods are as readily available for cats as they are for humans. Take a walk around any supermarket and you will see a vast array of canned, packaged or frozen options. There are foods for kittens, for adult cats and for mature cats. Many pet stores and veterinary clinics sell various 'professional formulae', some for ordinary maintenance and others for specific health problems.

When deciding whether to feed your cat a commercial diet or cook your own meals, there are various factors to consider. Many commercial diets are complete and balanced, which means that they will provide all the nutrients your cat needs. There is less certainty, though, that a homemade diet will be properly balanced.

You may also need to consider costs and convenience. Many commercial diets are more expensive than home-cooked ones, but home-cooked diets involve time, careful planning, preparation and storage.

Most cat owners find it convenient to feed reputable commercial diets and, if they wish, offer homemade 'treats' from time to time.

Pregnant queens, kittens, young growing cats and geriatric cats all have special dietary needs, so it is best to feed these a commercial diet specially formulated for their situation. Offer the occasional homemade meal to ring the changes.

Commercial diets

Many commercial diets are formulated to provide all your cat's nutritional requirements. They are formulated by nutritional scientists and veterinarians and have been tried and tested in controlled feeding trials and meet approved international standards.

There is a bewildering choice of commercial cat food. There are 'mainstream' foods, comparatively cheap, for the 'average' cat. There are 'premium' foods, often packaged in small quantities, that appeal to the human eye (and sometimes, but not always, to the cat's taste buds) and are more expensive. Your cat can have a choice of lamb, beef, chicken, tuna, sardines or ocean fish, to name but a few. Some foods contain combinations, such as beef and chicken. The nutritional profile of all these foods is very similar; it is just the contents that vary.

ABOVE: Like their human 'owners', cats love to eat, although unlike humans they are able to lose up to 40 per cent of their body weight without losing their lives.

NUTRITION
Commercial diets

Commercial cat diets can be grouped according to their moisture content:

- Canned or moist foods. These have a moisture content of around 78 per cent (roughly the same as fresh meat) and do not need preservatives because cooking destroys all bacteria and the canning prevents any further contamination. Because they contain no preservatives, if they are not used immediately after opening they require refrigeration.
- Semi-moist foods. These have a moisture content of around 30 per cent and normally contain preservatives. Some do not require refrigeration. They are commonly fed as 'treats'.
- Dry foods (complete diets). These have a moisture content of around 10 per cent, normally contain preservatives and do not require refrigeration. They are hygienic, easy to store and available for cats of all ages.

It is impossible to compare the relative nutritional and monetary value of all available products. To find out if a commercial food is fully balanced, check the label. There should be some statement to this effect, such as 'complete and balanced', and some labels bear a distinguishing mark that denotes that they have been tested and approved. Some formulas contain textured vegetable protein (TVP) as well as animal protein, because TVP is cheaper.

You will probably base your choice on a product's price, how readily your cat will eat it, and its labelled food content. The label usually lists the main food ingredients, and an analysis of certain nutrients such as protein, fat, and salt. Many manufacturers list the calorific value of the food, which can help you to decide how much to feed, and some indicate the amount that should be fed relative to body weight, stage of growth and activity level.

Foods from the veterinary clinic and pet stores

A number of international companies manufacture cat foods that are known as 'professional formulae'.

Available only from selected pet stores and most veterinary clinics, these differ from supermarket pet foods in that their constituents are guaranteed. What this means is that a cat food made of chicken, for example, will always contain a specified amount of chicken, no matter how high the cost of poultry at the time of production. The products also do not contain the TVP that is found in some supermarket products.

Other foods sold only through veterinary clinics are therapeutic diets formulated to assist in the management of health problems, such as allergies, gastrointestinal disorders, kidney and bladder disorders, liver disease and obesity. Special diets are available for pregnant and lactating queens, to assist cats convalescing after surgery or trauma, or for those undergoing treatment for conditions such as anaemia or cancer.

If you want information about any of these diets, talk to your veterinarian.

TOP LEFT AND RIGHT: Canned commercial diets contain balanced nutrients but do not help maintain healthy teeth and gums. Feeding dry food as well helps to overcome this problem.

Home-cooked diets

If you want to prepare some or all of your cat's meals yourself, and can ensure that it receives a properly balanced diet containing adequate amounts of animal protein, then by all means do so. If your cat has access to the outdoors, and to natural prey such as mice and lizards, then the chances are that the food it obtains outside will make up for any small deficiency in nutrients that might occur in your home-made diet. If it relies entirely on the food you provide, you must be absolutely sure that the diet you offer is properly balanced.

The animal protein for a home-cooked diet is usually derived from red meat, liver, kidney, heart, chicken, fish and (to a much lesser extent) milk. Remember that cooking food destroys some vitamins, and overcooking greatly reduces its nutritional value, so you need to supplement cooked food with the correct amounts and proportions of vitamins, just as reputable pet food manufacturers do.

Pet food supplements usually contain calcium carbonate or bonemeal (to create the crucial right balance of calcium and phosphorus), iodine, and vitamins A and D. You can purchase properly formulated supplements and various herbal preparations from a good pet store or some of the veterinary clinics.

Before basing your cat's diet on home-cooked foods and/or using any form of supplement, though, talk to your vet. Excessive supplementation with vitamins and minerals can cause serious health problems.

TOP: Cats enjoy eating prepared liver, although it is unwise to feed it to cats more than once a week. Large amounts of liver can give the cat too much vitamin A, which causes skeletal problems.

NUTRITION
Home-cooked diets

Supplementation of a home-cooked or commercial diet may sometimes be necessary for certain health conditions, such as stress, illness or post-operative recovery. In such cases you should always ask your veterinarian for advice, and he or she may recommend a change to one of the specially formulated therapeutic diets.

Ingredients for home-cooked meals
Even if you decide to make commercial cat foods the basis of your cat's diet, you may still find some of the information in this section useful.

Meat and meat by-products
All forms of red or white meat provide protein, B group vitamins, fat and energy, but the relative amounts depend on the type of meat and also on the cut.

Food	Protein (average %)	Fat (average %)	Energy (calories/100 g [3.5 oz])
Beef (medium fat)	20	15	220
Chicken (meat)	20	4.5	120
Chicken (necks)	13.2	18	230
Chicken (skin)	16	17	223
Lamb	15	22	265
Liver (ox)	20	3.8	140
Kidney (ox)	15	6.7	130
Heart (ox)	17	3.6	108

Chicken is considered to be more digestible than red meat. All types of meat and offal are seriously deficient in calcium and slightly deficient in phosphorus, and the proportion of phosphorus to calcium is greatly excessive, ranging from about 10:1 for rabbit and ox heart to 30:1 for veal and 360:1 for fresh liver. Meat is also deficient in vitamins A and D and iodine, copper, iron, magnesium and sodium, and needs to be supplemented with the missing nutrients, particularly calcium, if it is to form a balanced diet. Meat is most nutritious if fed raw, because cooking destroys much of its vitamin B content.

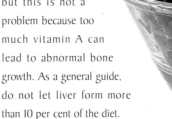

Liver is a valuable food rich in protein, fat, fat-soluble vitamins (A, D and E) and the B vitamins. Cooking reduces the liver's vitamin A content, but this is not a problem because too much vitamin A can lead to abnormal bone growth. As a general guide, do not let liver form more than 10 per cent of the diet.

Fish
There are two main types of fish. White fish has a nutrient composition similar to lean meat, contains less than two per cent fat, and is deficient in the fat-soluble vitamins (A, D, E and K).

Fatty and oily fish (such as tuna) contain high levels of vitamins A and D and high levels of unsaturated fatty acids, and feeding too much may cause a painful inflammation of fat deposits under the skin (steatitis).

Both white and oily types of fish contain high-quality protein and iodine, but are unfortunately deficient in calcium, phosphorus, copper, iron, magnesium and sodium.

Take care not to feed your cat too much raw (filleted) fish, as it contains thiaminase, an enzyme that destroys thiamine, one of the important B vitamins. Thiaminase is inactivated by heat, so it is best to cook fish before feeding.

Fish bones can cause problems if they get caught on a cat's teeth or stuck in its throat, so if you are feeding whole (unfilleted) fish, make sure that the bones have been softened by pressure cooking (this rather old-fashioned style of cooking is the ideal way to prepare fish bones for both cats and dogs), boiling or stewing. Whole fish fed in this way is nutritionally better than meat.

Eggs

Eggs contain iron, protein, most vitamins (except for vitamin C), fats and carbohydrates. Whole eggs contain about 13 per cent protein, 11.5 per cent fat, and provide about 160 kilocalories in every 100 g (3.5 oz). They are a well-balanced food and a useful source of animal protein and essential nutrients, particularly if fed raw. However, too much raw egg may be harmful, as egg white contains a substance called avidin that can reduce the availability of the B vitamin biotin, essential for many body processes including health of skin and hair, and proper muscle function. As a guide, feed no more than one raw egg per week to an adult cat. Cooking eggs by hard-boiling, poaching or frying reduces the avidin but also reduces their nutritional value, unfortunately. If you feed your cat the egg yolk only, you may increase the number of eggs to two or three per week. Remember that egg yolk on its own has a comparatively high fat content (about 31 per cent) and too much of this could cause obesity.

Milk, cheese and yoghurt

Dairy produce is high in protein, fat, carbohydrate, calcium, phosphorus, vitamin A and the B vitamins.

Whole milk is a useful source of calcium for kittens, and most cats like to drink it. You can serve it warmed, at room temperature or straight from the fridge – whichever way your cat prefers. Whole milk contains milk sugar (lactose), though, and as kittens mature their ability to digest this decreases. If fed more than small quantities of milk they may develop diarrhoea. Some adult cats are lactose intolerant, and if fed milk will develop an allergic, dry, itchy skin. For this reason low-lactose cat milk is manufactured and available in supermarkets.

Cream contains most of the milk fat and is a high source of energy, but if fed to excess it could result in obesity.

Cheese is a useful source of animal protein, and some cats like it. It does not contain lactose, so it can be fed in small chunks to cats that are known to have lactose intolerance. Pasteurized yoghurt also contains no lactose, but not all cats will consume it.

Fats and oils

A fat deficiency in the diet produces an itchy skin that may become dry and scurfy.

Fat is almost 100 per cent digestible and adds palatability to food – and, of course, cats love to eat it. Vegetable oils and fish fats are nutritionally better than animal fats. Safflower oil and corn oil are excellent sources of fatty acids – safflower is the best. If your cat's diet is not already balanced, you can feed very small amounts of cod liver oil (about a quarter of a teaspoon three times a week). Be very careful with such supplementation, and before using it talk to your veterinarian. Cod liver oil contains excessive amounts of unsaturated fatty acids and these may cause steatitis.

NUTRITION
Food groups

Vegetables
Most greens are rich in vitamin C, and vegetables are a good source of B-group vitamins. Cats can synthesize vitamin C in their body and don't require a dietary source. Some cats will eat greens and vegetables if part of a stew with meat or fish – remember that overcooking reduces their nutritional value.

Grains
Grains provide carbohydrate and some proteins, minerals and vitamins. They are generally deficient in fat, essential fatty acids and the fat-soluble vitamins A, D and E.

Wheatgerm contains thiamine and vitamin E. Wheat, oats and barley have a higher protein content and less fat than maize and rice. Rice is palatable to cats and is used as an ingredient in a number of commercial cat foods.

Yeast is rich in the B vitamins and some minerals, and yeast preparations may be beneficial to older cats – they are safe even if used to excess. Despite anecdotal evidence, dietary supplementation with yeast does not prevent fleas.

Fibre
Your cat's normal diet should contain about five per cent fibre (measured on a dry basis), derived from vegetable matter. Fibre-rich diets (10–15 per cent) may be used to help reduce obesity, and can also be used as a dietary aid in diabetic cats, because fibre slows the absorption of glucose (the end-product of carbohydrate digestion) following a meal.

Bones
Bones and bonemeal contain 30 per cent calcium and 15 per cent phosphorus, magnesium and proteins. They are deficient in fat, essential fatty acids and vitamins. Too many bones can cause constipation, so ask your vet about quantities.

It is also useful to check with your vet before you feed chicken bones. Rather do not feed mature chicken bones unless they have been softened by pressure cooking, because they are liable to splinter. Fish bones can get stuck in a cat's mouth or throat and should only be fed if they have been softened by cooking.

TOP: Cats are spoilt for choice with so many brands of cat food available. Some will develop a favourite, although most cats enjoy a variety.

Water

Make sure that clean, fresh water is always available. A cat's normal daily requirement (from feeding and drinking) is about 40 ml (about 2.5 tablespoons) per kg (2.2 lb) of body weight. Water intake will vary according to the environmental temperature and your cat's diet, increasing in proportion to the amount of dry food it consumes. It also increases if your cat is suffering from an ailment such as diarrhoea, diabetes or kidney disease.

Feeding

Select a feeding area in a cool place where your cat can eat without interference, and use it routinely. Use bowls made of easily cleaned materials such as stainless steel, earthenware or plastic, and wash them after each use.

Cats are known as fussy eaters, but with good reason. A cat will eat only the freshest food it can find, turning up its nose at the smell of anything old or stale. This is because cats are stimulated to eat not only by hunger, but by scent, too. As a general rule, they prefer to eat their food warm or at room temperature because it has a better aroma. Some will eat cold, unused canned food that has been refrigerated immediately after use, but others will not.

Because canned (moist) cat food does not contain any preservatives, remove any that is uneaten within an hour or so. If you do not have fly screens fitted, and flies are a problem inside your house, remove the uneaten food immediately. Semi-moist food can be left in a bowl for several hours.

If your cat does not overeat, dry foods can be left out all day. If you wish to do this, purchase a proper self-feeder that keeps the food (and its aroma) enclosed and allows the cat free access.

Feeding your kitten

A queen's milk is rich in protein and fat, and during the first few weeks after weaning, a kitten's diet needs to reflect this. A growing kitten requires up to three times more energy intake per kg (2.2 lb) of body weight than does an adult, and because it has a limited stomach capacity it must be fed several times a day on a high-energy diet.

ABOVE: No matter how well fed, cats will succumb to instinctive behaviour and prey on birds and small mammals, should the opportunity arise.

NUTRITION
Nutritional problems

There are many commercial brands of food specially formulated for kittens, both grain-based and meat-based, which you should consider rather than trying to formulate your own, homemade diet.

When a kitten is young you can also feed milk – although many vets do not recommend that you feed cow's milk to either a kitten or a cat. If feeding cow's milk appears to cause diarrhoea as your kitten gets older, it could be because of the milk's lactose content, so change to a specially formulated cat milk.

As a general guide, a kitten aged 8–12 weeks should receive at least four meals a day of a commercial food or a home-cooked diet. You need to decide which you are going to use, and stick to it. Mixing home-cooked and commercial diets can lead to imbalances.

From three to six months feed at least three meals a day, and introduce the regime suggested below for adult cats.

Feeding your adult cat

Unlike dogs, which are competitive pack animals and will happily wolf down as much food as their stomach can handle, cats are solitary hunters and don't usually eat a large amount at a time. Given the option they prefer to eat their daily ration in several small meals.

Most owners find it convenient to feed their cats a mixture of moist and dry foods. Dry foods can be left out for a cat to self-feed and moist foods can be fed in controlled amounts two or three times a day in the morning, early afternoon or early evening.

Late-night feeding can cause problems if your cat is confined to the house during the night, because most cats need to urinate and defecate within an hour or two of feeding.

If you are feeding more than one cat, you may need to feed them separately and some distance apart. This way a dominant cat cannot eat another's food, and you can monitor their individual food intake.

How much to feed

Most cats tend to eat only enough to satisfy their energy needs. The amount of energy your cat uses will depend not only on its activity but also on its metabolic rate (the speed at which it burns up the energy). Every cat is an individual, and there can be as much as 20 per cent variation between two similar cats.

Your cat should be fed enough food to satisfy its energy needs, but no more – otherwise it will put on excess weight. Excess energy is stored as fat, deposited under the skin and under the abdomen (causing an appearance referred to as an 'apron'). Some commercial diets are particularly palatable, and stimulate a cat to overeat. If you feed a commercial diet, ascertain its energy content from the label and feed your cat accordingly. If you feed a homemade diet, it may be more difficult to determine exactly how much to feed, and you will need to closely monitor your cat's health.

The most important criteria for judging if you are feeding the correct quantity and balance are the health and appearance of your cat. If it is in good condition, alert and active with a healthy skin and coat, and maintaining its proper weight, it is almost certainly getting an adequate diet. If it has a scurfy skin, is shedding its coat excessively, is over- or underweight, appears dull or listless, is excessively hungry or often disinterested in food, you should talk to your vet.

Remember that if you give your cat treats between meals, these contain calories and you must make allowances for them in your cat's overall diet.

Nutritional problems

These should not arise if you are feeding a properly formulated commercial diet. They may arise because a cat is:

- receiving the wrong diet
- eating, but a disease is reducing its ability to absorb or use food
- not eating for a variety of reasons.

Underfeeding results in lack of energy, weight loss (the body burns up fat reserves, then the protein in the muscles) and finally starvation. It could also result in a deficiency of essential nutrients.

Overfeeding causes obesity and perhaps a toxicity caused by a nutrient excess (such as vitamin A).

UNDERSTANDING YOUR CAT
The feline social system

The social system of cats living in a wild state varies in relation to their ecological circumstances – food availability is the major influencing factor. Groups may be formed when the availability and dispersion of food allows two or more individuals to live in close proximity. Most of these groups consist of females, usually related, together with their offspring and immature males.

Females often nurse each other's kittens and bring back prey for them all. Mature males are not part of the 'family' group, forming only loose, temporary liaisons with it for breeding purposes. They are not involved with the rearing of kittens.

Where food is scattered, cats live a more solitary existence. Territories are formed and defended areas contain the denning site and possibly a major food source. These are marked using scent and visual signals, such as scratch marks (claw-marking) and uncovered faeces. Cats entering the territory of others risk attack, although home ranges

ABOVE: Given the opportunity, cats will raise their young communally in a similar way to lions. Several females, usually sisters, will help with the kittens and take turns babysitting while the others hunt.
TOP: Cats enjoy high places, which is why they favour windows as entry and exit points.

(areas in which hunting is done) may overlap. However, where this occurs, the cats sharing the home range rarely meet. They seem to have some sort of 'time-share' arrangement, which ensures that they hunt in different areas at different times.

The cat-human relationship

Cats are flexible in their dependence on humans, as most cats are capable of surviving in a wild environment. In suburban areas it is difficult to find enough prey to support a totally feral existence, so a liaison with humans is necessary and beneficial. The relationship between cats and humans is one of mutual gain or symbiosis. Cats gain shelter, a food supply and health care. We get rodent control and companionship.

Unlike dogs, cats do not necessarily regard humans as part of their own social group. The social order of a group of cats in a household tends to involve only the cats. This does vary, though: cats raised in close contact with humans, and who have no other cats in the household, do regard humans as part of their social group, and may show behaviour such as status-related aggression toward them (see p61). Some breeds, such as Siamese and Burmese, have been selectively bred for friendliness toward humans and are often very attached to and psychologically dependent on their owners. On the whole, most cats maintain a relatively independent existence and seek human companionship on their own terms.

A kitten's social development

Kittens are born blind and deaf. Their eyes open at two weeks and they begin to play at three. At this age they hear well and have a good sense of smell, and the gape response is shown (see p58). Vision develops slowly, over 10 weeks.

ABOVE: Kittens are born blind and deaf and are relatively helpless before two weeks of age.
TOP: It is essential that children are taught to hold a cat correctly, with its weight supported by the holder's hands or body. An uncomfortable cat will kick and bite in an attempt to free itself.

Smell is probably the most important of the cat's senses

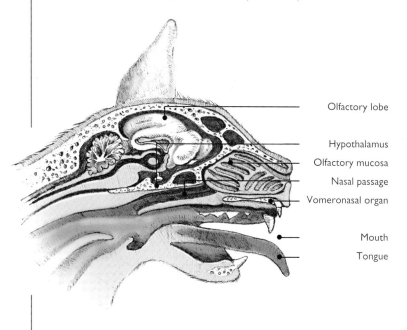

Cats are able to detect tiny particles of scent, which tell them about the social and reproductive status of other cats. They do this by means of the 'gape' response (also known as 'flehming') – this lip-curling expression delivers the chemicals of the particular scent to the sensitive vomeronasal organ.

Play increases in intensity from four to 11 weeks, then declines. By eight weeks the kittens are capable of killing and eating small prey. It is important that kittens socialize with humans and other cats during the two- to seven-week period, otherwise they will be fearful of human contact. If hand-reared and isolated from other cats or kittens, they may never relate properly to other cats.

The cat's sense of smell

Cats have a keen sense of smell and have a special structure (the vomeronasal organ) that helps them detect scents. This organ consists of two blind-ending tubes that run between the oral and nasal cavities. Substances taken into the mouth are dissolved in fluid contained in the fine tubes, then sensory information is conveyed to the olfactory organs.

When cats are investigating a smell using this organ, they hold their mouths open in a sort of gaping grimace.

Scent is an important means of communication. It is used to define territory via sprayed urine or from scent deposited from facial glands during rubbing or glands in the feet during scratching. Scent is also deposited onto faeces from the anal glands. It is thought that scent conveys information about identity, social status and the reproductive state of the cat. There is also evidence for the existence of clan odours, identifying members of a group of related animals.

Scent or odour is also important in stimulating appetite in cats and in prey identification. Cats with upper respiratory tract virus infections that cause nasal congestion often will not eat, as they cannot smell the food.

UNDERSTANDING YOUR CAT
Your cat's senses

Vision

Cats have good stereoscopic vision and are excellent at detecting movement. Their ability to detect light is three to eight times better than that of humans. Their visual range extends from 25 cm (10 in) to 2 m (6.5 ft), and they have excellent depth perception and blue-green colour vision.

Cats are able to see well in the dark, thanks to a special structure (the *tapetum cellulosum*) that reflects light back onto the retina.

Many Siamese cats have poor stereoscopic vision, and reduced close vision and flicker detection as a result of an inherited defect. Many of these cats have obvious squints.

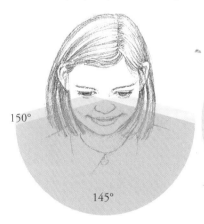

A human sees from side-to-side – a total of 150° – of which 145° is binocular overlap.

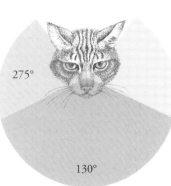

A cat sees from side-to-side – a total of 275° – of which 130° is binocular overlap.

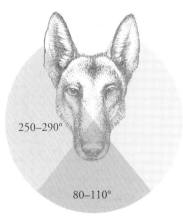

A dog has a total vision of 250–290°, with 80–110° of overlap – much less than that of humans.

TOP: Cats have excellent night vision, which is partly due to the *tapetum cellulosum*, which reflects light back onto the retina.

Sound

Cats are very sensitive to high frequency sound (60 khz), which enables them to detect rodents' ultrasonic squeaks. They are good at judging the height of a sound's origin, and the mobile external ear (pinna) helps in sound localization.

In addition to their acute hearing, cats have vibration detectors in their feet, which are able to detect 200 to 400 hz – but only for short periods of time.

Vocalization

Cats scream when attacked or frightened, yowl and wail menacingly when warning off intruders, miaow loudly for attention and chirrup a greeting to familiar animals and people.

Purring usually occurs during pleasant experiences, such as suckling or being stroked, although it is sometimes done during times of stress or severe illness. Cats never purr when they are asleep.

Like we do, cats give away their 'feelings' by the expressions on their faces

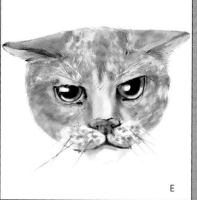

A: A happy cat – the ears are upright and the whiskers relaxed.
B: Quite content – eyes half closed and whiskers relaxed, this cat is probably being stroked and patted.
C: On the alert – the cat takes on an expression somewhere between content and nervous.
D: Feeling nervous – the ears begin to move back and the whiskers move slightly forward.
E: Angry and frightened – the ears are flat, the eyes are narrow and the whiskers are forward.

TOP: This cat's confident body posture with an upright tail indicates a willingness to 'communicate'.

UNDERSTANDING YOUR CAT
Behaviour problems

Behaviour problems

The feline socialization system is very effective, and cats rarely exhibit seriously problematic or aggressive behaviour, unless they have not been de-sexed or are in overcrowded conditions.

Nevertheless, where cats do exhibit behaviour problems, it is usually possible to deal with them effectively.

Status-related aggression

Some cats are dominant by nature and need to feel in control. They will often growl at or bite owners for no apparent reason. Often this happens while they are sitting on the owner's lap, being patted. They will suddenly tense up, their pupils will dilate, their tail may lash and they will bite or take hold of the owner's hand in their teeth.

These cats are also likely to attack if the owner has been patting them and then decides to place them on a chair and move, or just decides to move position while still sitting with the cat.

Dealing with status-related aggression

Watch for the signs of impending aggression. Stop patting the cat as soon as it appears tense, stand up without touching it and let it fall to the floor. A water pistol or foghorn may be used to startle it out of attack mode.

Preventing status-related aggression

There is no way to prevent status-related aggression from developing. You will have to accept that these cats will never be cuddly – they will always need to be in control.

Redirected aggression

Cats are masters at 'taking it out' on others. If a cat sees another intruding onto its territory but cannot access it to drive it away (for example, if it is inside and sees the intruder through a window), it will redirect its aggression onto whoever is close to it – very often you!

If you approach the cat while it is growling at the cat intruder, it is likely to attack you.

TOP: An arched back and fur standing on end indicate fear or aggression. This is designed to make the cat seem larger and therefore more intimidating. Kittens practise these postures in play.

Redirected aggression can also happen if the cat is frightened; if something falls off a shelf and scares it, and at the same time you enter the room, the cat may associate you with the frightening experience and attack you.

Dealing with redirected aggression
The best thing to do is to walk away and leave the cat to calm down. Do not attempt to approach or pacify the cat, as it will remain reactive for a long time in these situations. If the cat continues to react negatively towards you, you may need the help of a behaviourist to overcome the problem. Anti-anxiety medication may be required, along with a behaviour modification programme.

Preventing redirected aggression
Avoid approaching your cat if it is reacting to something outside. Never rush to comfort your cat if something has fallen near it or if it has had a major fright for any reason.

Predatory aggression
Some cats will attack people as if they were prey – they stalk and attack quite savagely. The behaviour is especially dangerous when directed toward elderly people or children.

Dealing with predatory aggression
When the cat begins to stalk, startle it using a foghorn or a water pistol. Try to avoid situations that are known to provoke it, such as wiggling toes inside socks or sandals or children wearing dangling ribbons or untied belts.

Some cats will lie in wait for owners coming to breakfast or following a regular routine. Try to vary your routine so that the cat cannot predict where and when you may pass by.

Preventing predatory aggression
Do not encourage predatory behaviour in kittens – for example, do not play games of 'hide and seek' or encourage your kitten to chase you.

ABOVE: This kitten is preparing to pounce on a playmate – a posture also seen in hunting.

UNDERSTANDING YOUR CAT
Aggression

Place a bell on the cat's collar so that you may be warned of its approach. (This is not infallible, though, as cats can learn to move without ringing the bell.)

Play aggression

This is seen most often in hand-reared kittens. The lack of contact with littermates means that they do not learn to inhibit their biting and sheath their claws while playing. These cats can bite hard enough to draw blood, although they have no intention to injure.

Dealing with play aggression

When the kitten or cat starts to elicit play, by batting at your hand or jumping at your feet, redirect the attack onto a toy. If the cat insists on body contact, use a water pistol to discourage rough play. Squirt the cat in the face and say 'ouch'.

Reward quiet, gentle play with treats such as pieces of cheese or cat biscuits. Never encourage these cats to chase hands or toes, even if these are hidden beneath blankets, as they will probably bite straight through.

Preventing play aggression

Ensure that young kittens are socialized with other kittens or adult cats. Discourage play involving human body parts.

Fear aggression

Cats or kittens that have not been exposed to humans early in life are often fearful of people and may be aggressive when approached. Such a cat will back away, crouch or lie over on one side and hiss with ears flattened and pupils dilated. It will attempt to escape contact, but will bite if this is prevented.

This sort of behaviour is often seen in kittens from wild colonies that people decide to adopt, and in cats that have

belonged to one elderly owner who has died and the cat is being re-homed. Cats may also show fear aggression when taken out of their home environment, to a vet clinic or cattery.

Dealing with fear aggression

With wild-born kittens or re-homed isolated cats, fear aggression can usually be overcome with time and patience. However, some wild-born kittens are genetically fearful and may never be cuddly pets.

ABOVE: Fear aggression is shown by this cat's crouched posture and flattened ears.
TOP: Kittens develop physical skills through play and practise techniques required for self-defence in later life.

Keep the cat in a room with a litter tray, a bed, a dish of water and plenty of toys. Spend time in several sessions every day just sitting in the room beside the cat or reading. Reading aloud often helps. Take tasty, strong-smelling cat food with you and attempt to hand-feed the cat. At first you will probably just have to drop the food nearby, but gradually the cat should move closer and closer until the food is taken directly from the hand. Avoid eye contact at this stage.

Do not attempt to touch the cat until it is relaxed when eating from your hand. Gradually entice the cat to play, using paper and string or a flexible stick. Most cats will respond within two to four weeks. Once they can be petted and are relaxed in your company, they can be allowed access to the rest of the house and finally to the garden.

Do not attempt to pick up these cats until they are very relaxed, which will probably not be before three months after adoption. In severe cases, they may need anti-anxiety medication.

With pet cats in strange surroundings, it is best to try to desensitize the cat to the whole experience.

Firstly, ensure that the cat is comfortable with its carry cage. Feed it in the cage from time to time, and reward it for allowing you to place it in and remove it from the cage. If your cat responds to the herb catnip, place some in the cage. Play with the cat's favourite toy in and out of the cage.

If the cat is fearful of the vet clinic, arrange with your vet to bring the cat in regularly and allow it to investigate the rooms and be fed there. If the cattery is a problem, see if the owners will allow you to do the same thing there. Most cats will eventually learn to tolerate, if not enjoy these necessary visits.

Aggression towards other cats

Cats are territorial by nature, and will naturally threaten and drive off intruders. It is not really possible to modify this behaviour, but you can reduce the chances of your cat being involved in territorial disputes by keeping it indoors at night and in the early morning and late evening, as this is when most disputes are likely to occur.

Aggression towards other cats in the household

Cats living together in a household frequently have minor disputes that are settled by a hiss and a swat with a paw. Usually they can sort it out and will at least form a truce, if not a loving friendship.

Some cats do become very attached to each other, and wash and groom each other and sleep together. In other cases, though, there is constant and extreme aggression shown towards a cat by one of its housemates. This may or may not be associated with defence of territory (an area that the cat defends as its own). It can also result from redirected aggression.

ABOVE: These two cats are involved in a typical aggressive encounter. Note how the cat on the left is more dominant – its ears are less flattened and it has a more upright posture.

UNDERSTANDING YOUR CAT
Inter-cat aggression

Dealing with inter-cat aggression

Where cats are showing extreme aggression, it is necessary to separate them. Confine them to separate rooms, and swap them daily so that they remain in contact with the scent of the other cat. Teach them both to accept being in a carry cage and feed them in this. Play with each cat at a set time every day, and offer treats.

After a week, bring the cats out at feed time, in their cages. Place them at opposite ends of a room (not one of the rooms in which they have been kept) and feed them both in their cages. At the first sign of reactivity, remove the offender from the room. Once the cats accept being caged at a distance in the same room, gradually move the cages closer. Once they can quietly sit and eat at a distance of 2–3 m (5–10 ft), feed them out of their cages.

If all goes well, then place them both on harnesses and leads and sit in the room with them at a distance from each other. Any sign of aggression should be reprimanded using a foghorn or a water pistol, and calm behaviour should be rewarded. If they are coping, start a series of play sessions using their favourite toys and involving both cats. Eventually they will be able to remain free in the room together, preferably with toys and definitely under supervision. Any aggression should be instantly reprimanded. With time the cats should be able to coexist happily or at least without constant fighting.

Dealing with redirected inter-cat aggression

Redirection of aggression towards other cats occurs in the same way as described for redirection towards people (see pp61–2). The difference is that the cat against whom the aggression is redirected tends to become terrified of the aggressor. When the two cats meet, the victim shows signs of fear, which reinforces the aggressive response in the other cat. These cats should be dealt with as described for inter-cat aggression, although the victim will probably need anti-anxiety medication to stop it from behaving in a way that triggers aggression in the other cat. The aggressor sometimes also needs medication. This situation can be difficult to deal with and the advice of an animal behaviourist should be sought.

Fabric eating

Most cats are fastidious in their eating habits, and are reluctant even to try new foods. Some cats, however, can develop bizarre tastes, and the most common of these is for fabric.

Fabric eating is most frequently seen in Siamese cats – it seems that there is a genetic predisposition. The behaviour has also been associated with early weaning, and may be triggered by a traumatic event such as moving house.

ABOVE: Kittens love to play with wool and fabric, but unfortunately some develop a habit of eating the material as well. This can cause digestive upsets and should be prevented.

Fabric-eating cats often steal woollen clothing from neighbours, drag it home and suck and chew at it. They may eat large amounts and as a result can suffer digestive upsets and blockages.

Dealing with fabric eating

Try to prevent any access to woollen fabric, and perhaps encourage these cats to rip up cardboard. You could also give them bones or rawhide to chew on.

Some cats stop eating fabric if you use aversive substances such as chilli on the fabric. It also helps to increase the amount of attention you give to your cat, and provide more stimulation such as games and new toys.

Some vets believe that these cats may suffer from an underlying neuro-chemical abnormality, so medication may be required.

Preventing fabric eating

Try to avoid using woollen bedding for kittens. Provide plenty of stimulation and activity and avoid early weaning of kittens. Avoid purchasing kittens from fabric-eating parents.

House soiling

Many people experience problems with house soiling during their cat-owning lives. There are several possible causes.

Any cat living inside should have access to a litter tray. If it is free to enter and leave the house at will, it may never use one, but if the weather is cold and wet, or for some reason it feels apprehensive about venturing outside, it is useful to have a litter tray available.

In multi-cat households there should be one litter tray per cat, and one extra. These boxes should be changed daily if soiled. Types of litter include bark, sand, sawdust, recycled paper pellets and granulated absorbent pellets. Perfumed substances are sometimes available.

Litter trays should be big enough for the cat to turn around in comfortably, and contain sufficient depth of litter to allow digging. They should also be stable, as if they wobble the cat will feel insecure.

Soiling near the litter tray

This is usually a problem related to the type of litter in the tray, or to an association of pain or unpleasantness with using the litter tray. Often the paper around the tray will be scratched up in an attempt to dig and cover.

Dealing with soiling near the litter tray

Check that no other cat has been using the box, as most cats resent sharing, and ensure that the box is kept clean and fresh. If there is no apparent problem, have the cat checked by a veterinarian. It may have a bladder or bowel problem that has caused it to anticipate pain with getting into the box. If it is an old cat it may have arthritis, making it uncomfortable to climb into the box and balance.

If the cat is healthy, consider changing the type of litter. Offer the cat several types of litter in different boxes and see which one, if any, it uses. Place the boxes in different areas to which the cat has ready access and in which you would be happy for the litter box to stay permanently if need be. Rotate the litter types so that you have tried all types in all possible areas. The reason for this is that it may not be a litter problem – it may be that the cat has been chased by another cat or otherwise disturbed while digging in the box and so is afraid to climb into it in that area. If you cannot solve the problem, consult an animal behaviourist.

TOP: Litter or sand trays should be clean, roomy and placed in an area allowing some privacy.

Preventing soiling near the litter tray

Ensure that there are always enough clean litter trays for each cat, plus one extra.

Urination in the house

Cats may urinate in other parts of the house because:

- they have cystitis (a bladder infection)
- they have a litter aversion (see p66)
- they are afraid to use the litter tray because it is in a high-traffic area
- another cat has used the tray or another cat is intimidating them when they try to approach the tray
- they are marking part of the house as their territory.

Urine may be sprayed or deposited from a squat.

Spraying is not usually associated with ill health. It is usually marking territory or a way of expressing aggression by cats that do not have the confidence to engage in direct conflict. Cats that are spraying or urinating in the house are often very anxious.

Dealing with urination in the house

Take the cat to the vet for a check-up. Confine the cat to one easily cleaned room when you cannot supervise it, and provide a litter tray. If the cat doesn't use the tray while in the room, treat as described for soiling near the litter tray.

If the cat is using the litter tray in the room, gradually reintroduce it to the rest of the house, having cleaned all previously soiled areas with an enzymatic cleaner. Do not allow the cat to be anywhere unsupervised. If it attempts to dig, squat or spray, squirt it with a water pistol or alarm it with a foghorn, but do not physically or verbally abuse it.

If the cat is marking territory, it may be that the social relationships between the cats in the household have somehow changed. Marking is often seen when a new cat is introduced to the household. In a single cat household, marking may be done in response to a visitor staying for a few days or a new partner moving in. It can also occur in response to the presence of cats outside the house, and as a reaction to strange cats entering the house. Try to identify any major triggering factors. For example, if the cat is urinating or spraying on the windowsill it is probably reacting to the presence of a strange cat outside. Fitting blinds to the window and keeping the cat out of the room when these are not closed may solve the problem. If you have seen strange cats hanging around, invest in a cat door that is electronically operated by a trigger on your cat's collar. This ensures that no other cat can enter.

If you feel that the cat may be stressed by changes that have occurred within the household, try to provide it with extra attention and some stability by playing a game with it at a fixed time daily. This helps to reduce anxiety.

These problems can be difficult to solve and you may need the help of an animal behaviourist to correctly identify the cause of the problem.

Preventing urination in the house

Ensure that strange cats cannot enter the premises. Provide adequate numbers of clean litter trays. Try not to change the home environment suddenly. Introduce the cat to new flatmates before they move in and take things slowly. Take care to introduce new cats or other new pets to the cat gradually, too. Provide some routines and ensure that each cat has a special place to which it can retreat such as a shelf on a cat tree, cupboard or cardboard box.

Defecating in the house

Cats use faeces as a territorial marker – this is known as middening. Faeces are coated with a secretion from two glands situated on either side of the anus, which contains information about the cat which other cats will receive when they sniff at the faeces. Where disputes are happening it is not uncommon for cats to use faeces as well as urine to make a point. Other reasons for defecating in the house include litter aversion, ill health and poor house training.

Dealing with and preventing defecation in the house

Follow the same protocol as for urination in the house. There may be underlying stress factors that need to be dealt with, and an animal behaviourist should be consulted if the problem continues.

Hunting

Cats are natural predators. Even kittens that are hand-reared in total isolation may still grow up to be effective hunters. Prey includes birds, rodents and lizards. It is not uncommon for cats to bring their prey home to present to their owner. Many people find predation by their cats unacceptable. This is especially true in areas where cats may prey on endangered wildlife.

Dealing with hunting

Since cats do most of their hunting early in the morning, in the evening or late at night, keeping them confined at these times will greatly reduce their predatory behaviour.

The only way to completely prevent it is to keep your cat confined inside continuously. Cats can and do adapt to this lifestyle, provided they are provided with plenty of stimulation.

Other methods of reducing predation include placing a collar and bell on the cat and the application of startle tactics such as foghorns whenever the stalking cat is observed. These methods are not very successful, though. Cats soon learn to position themselves in such a way that bells will not sound during a stalk, and although startle tactics may prevent a cat from targeting the bird table they will do nothing to reduce predation in other areas when it is far away and out of sight.

'Liberator' collars produce a series of beeps in response to a sudden leap by the cat, but have not proved to be completely successful either.

Ensuring that cats are well fed before going out may reduce the amount of predation slightly. However cats are still likely to hunt when well fed – they just won't eat the prey and may play with it for longer prior to killing it.

Preventing hunting

As this is instinctive behaviour, cats cannot be taught not to hunt. The only way to prevent hunting, unfortunately, is to deny access to prey.

Jumping onto high surfaces

Cats enjoy vertical space. Most love to climb and to be up high. Benches, shelves, tables, fridge tops and mantelpieces are appealing to cats for this reason. Not only are they able to survey their territory from a vantage point, but they are also able to satisfy their curiosity about everything. This is not something that most owners welcome.

Dealing with jumping onto high surfaces

Cats can be trained not to jump onto forbidden surfaces. Try squirting them with a water pistol as soon as they jump up, or startling them with a loud noise. Placing something sticky all over the surface (try honey) may act as a deterrent. A smooth piece of plastic or card that slips easily and will fall off when the cat hits the surface can be very effective, too.

TOP: Try not to apply 'human morality' to cats and regard their instinctive hunting tendencies as cruel – cats are simply honing their inherited hunting skills.

UNDERSTANDING YOUR CAT

Hunting, jumping and scratching

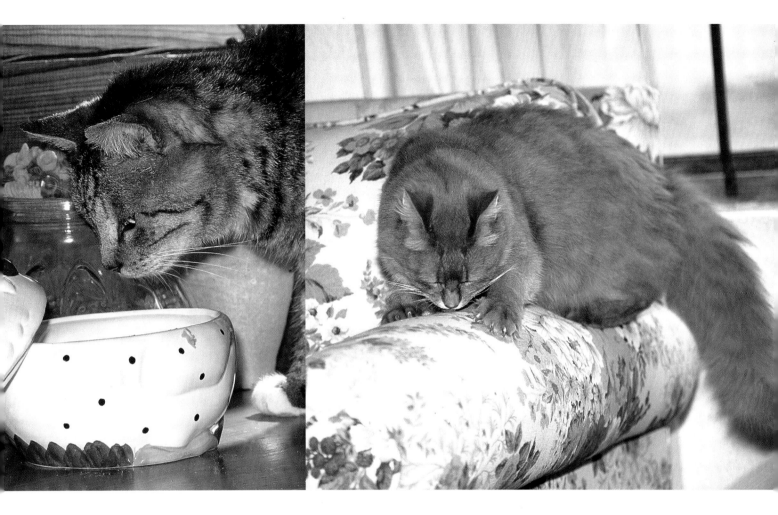

Preventing jumping onto high surfaces
- Train your cat from kittenhood.
- Don't tantalize your kitten or cat by preparing its food on the counter while it watches hopefully from the floor or nearby chair. Prepare the food before the cat arrives, ready to place in its feeding area.
- Provide your cat with vertical space of its own, such as a cat tree or a high shelf with its own special bed from which to view the world.

Scratching furniture

Cats scratch as a means of marking territory. The marks provide a visual signal, and scent from the sebaceous glands in the feet is deposited during the scratching process. Favoured sites are usually vertical wooden surfaces. Outside, tree trunks and fence posts are often used. Inside, furniture will do nicely.

Dealing with furniture-scratching

Cats often like to scratch as soon as they wake up, so it is a good idea to provide a scratching post near to your cat's sleeping area.

Scratching posts come in a variety of shapes, sizes and materials – cats seem to enjoy the ones made of wood and covered with carpet. Others combine a serrated dense card which can be shredded, or use hessian as a covering.

If your cat insists on using the furniture as a scratching post, there are two main options: either train the cat not to scratch except in certain approved areas, or protect the furniture by decreasing the amount of damage that the claws

TOP LEFT: Cats are naturally curious and love to explore table tops, especially if they might find food there.
TOP RIGHT: Cats prefer to scratch their claws against an upright feature, but if none is available, the horizontal arms of your furniture will do.

may do. A cat's claws may be trimmed in much the same way as a dog's toenails, and if this is done regularly from kittenhood cats accept the procedure quite happily. Ask your vet to demonstrate how this is done. You will need to trim the claws every four to six weeks. An alternative to this is to apply plastic sheaths to the claws – these are glued on much the same as false fingernails.

To train your cat not to scratch furniture you need to be vigilant, and the cat must not have unsupervised access to the house. Keep a water pistol handy and squirt the cat as soon as it attempts to scratch, or use a foghorn to startle it. Other methods include booby-trapping areas with small balloons, which will pop when scratched and alarm the cat.

Preventing furniture-scratching

Encourage your kitten to use a scratching post – suspend a toy from the post, or place catnip around the post as an incentive (not all cats are susceptible to catnip, though).

Declawing

This must be regarded as an absolute last resort. Declawing is the surgical amputation of the first joint of every digit on the cat's forefeet. It is a mutilation that causes much post-operative pain, and it is possible that declawed cats experience phantom pain during life.

This procedure is illegal in some countries, for example the United Kingdom.

TOP: Cats need vertical space, and if they are not provided with climbing apparatus such as 'cat trees' or a garden, they will often resort to climbing the curtains.

UNDERSTANDING YOUR CAT

Separation anxiety

Declawing should never be necessary if owners are prepared to put the time and effort into training their cat not to claw furniture and regularly trimming the claws.

Yowling and crying

Some cats can be very vocal and seem to 'talk' incessantly. The Oriental breeds are more likely to behave in this way than other breeds. These cats become very attached to their owners and need company, and they are naturally vocal.

Unspayed females of any breed will yowl and cry when in season, and un-neutered males have a special yowl when seeking a mate.

Where vocalization has become excessive or has suddenly started in a previously quiet cat, there may be an associated health problem such as hyperthyroidism or, in an elderly cat, cognitive dysfunction or 'feline Alzheimer's' disease. If the cat is otherwise healthy, it could be a sign of anxiety. If the cat yowls when out of sight of the owner but is quiet in the owner's presence, it could be suffering from separation anxiety (see at right).

Dealing with yowling and crying

If the problem has started in a previously quiet cat, take it for a complete health check.

If all is well, consider if there have been any major changes in your cat's surroundings or lifestyle that may have upset the animal.

If the cat is yowling and rubbing around you but is quiet when you are not there, it is seeking attention. Try to reserve a special time to spend with the cat each day when it has your undivided attention, is given treats and is played with. At other times ignore the yowling and pet the cat when it is finally quiet. Provide plenty of interest for the cat in the form of toys for times when it is alone.

If there is no improvement as a result of these tactics, consult an animal behaviourist.

Preventing yowling and crying

- Ensure that food is freely available from a cat café (automatic feeder) throughout the day.
- Take turns at feeding the cat and playing with it.
- Install a cat door so that your cat does not have to demand to be let inside.
- Don't rush to the cat whenever it cries.
- Give your cat quality time with you every day.
- If you are a busy person, consider keeping two cats as company for each other. It is best to get two kittens rather than introduce another cat or kitten to an adult animal.

Separation anxiety

Cats may become extremely attached to their owners to the point that they are unhappy if left alone. These cats need to have constant owner contact. They cry incessantly when left, even if the owner is in an adjacent room. The condition is seen most commonly in Oriental cats and cats that have been hand-reared.

Dealing with separation anxiety

- Encourage the cat to relate to more than one family member.
- When the cat must be left alone, provide lots of environmental stimulation such as new toys and meaty bones to chew.
- Do not give in to demands for constant attention; choose when you wish to pat the cat and ignore it at other times.
- Create as much stability and routine in the cat's life as possible.
- In extreme cases, anti-anxiety medication may be required.

Preventing separation anxiety

- Ensure that all family members have contact with the cat.
- Do not give in to attention-seeking behaviour, supply affection on your terms.
- Do not carry hand-reared kittens around continuously, but give them time alone with a stuffed toy and hot water bottle.
- Some cats will show this behaviour no matter how they are brought up.

BREEDING AND REPRODUCTION
A new member of the family

Kittens are cute and delightful, but before deciding to let your mixed-breed cat have a litter, ask yourself how likely it is that you will find good homes for them. Remember that every year thousands of unwanted kittens are destroyed by humane societies.

There is no evidence to suggest that having a litter is physically or psychologically beneficial to your female cat, so don't feel that she is being deprived if she is spayed when she is six months old. If you have six to eight friends waiting eagerly for kittens from your cat, then you can probably allow her to breed with a clear conscience. If your cat is pedigree and you breed her to a pedigree tom, you should be able to place the kittens.

Ask the breeder of your cat or the owner of the sire if they have lists of people waiting for kittens.

If you do decide to breed from your cat, try to delay the event until she is at least a year old. Although cats can breed at six months, they do better if allowed to mature first.

When your cat first comes into season (also called 'on heat') you will notice a distinct change in her behaviour. She will become excessively friendly, roll around on the ground and yowl in a tone that you have never heard from her before. When rubbed on the back she will raise her hindquarters and tread with her hind limbs.

If your cat is not pedigree and is at the right age to be bred, all you have to do is let her go outside once she is on heat, and a roaming tom cat will quickly find her.

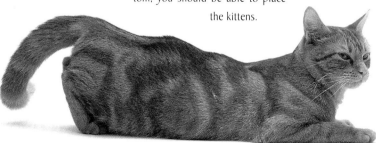

ABOVE LEFT: This body posture, known as 'lordosis', is typical of a female cat in season.
ABOVE RIGHT: Cats that are in season may scent-mark objects in the home by spraying them with urine or rubbing against them.
TOP: While mixed-breed kittens are as charming as pedigree kittens, they are more difficult to sell or give away, so think carefully before allowing your mixed-breed queen to mate.

Before you do this, ensure that she has had all her vaccinations and has been wormed and treated for fleas.

Cats are induced ovulators, which means that mating triggers ovulation (shedding of the egg from the ovary). They often mate with several toms during their season and as a result the kittens may have different fathers. Once a female has been successfully mated she will stop showing signs of heat.

If your cat is pedigree, you will need to keep her indoors, because tom cats from miles around will camp on your lawn and fight under your windows. Other bad news is that your female will come into season every three weeks from late winter/early spring until she is mated.

To arrange for a suitable mate for her, consult the breeder of your cat or contact your local Cat Fancy Association. Try to view the tom before sending your cat to him. Make sure the breeder is registered and that the tom is certified free from disease and is fully vaccinated. It is customary to send the queen to the tom because the tom may be distracted in unfamiliar surroundings, and some queens will attack toms that are suddenly introduced to their territory.

Gestation (time from mating to birth of kittens) lasts for 56 to 63 days. During this time the cat should be fed good-quality food on demand. Commercial foods are available for pregnant and lactating cats, and it is probably easiest to use

ABOVE: Pregnant queens usually remain agile and playful until they are close to giving birth.
TOP: The typical mating sequence. The male holds the female's neck, and as he dismounts, the female will usually turn and spit or hiss at him.

these. If you choose to feed a home-prepared meat diet, you will need to provide a calcium supplement. See your vet for advice on this. Your cat should also be wormed monthly and treated for fleas with a topical treatment that will not harm the kittens (check with your vet).

During your cat's pregnancy, get her accustomed to the area in which you would prefer her to give birth. This may be a specially prepared box or basket in a spare room or the bottom of a wardrobe. She needs a place where she can feel secure and relaxed, and it should also be easy to keep warm and clean. It is helpful to have her sleeping in this area throughout her pregnancy.

Giving birth

About 12 hours before giving birth your cat will seem restless and agitated. She may eat less than usual, and may vocalize and seek human company. Encourage her to stay in the kittening area, and if necessary sit with her as time permits and encourage her to relax.

Eventually the birth contractions will start, and at this stage most cats are best left alone. Check on her every 20 minutes, or allow one person to remain in the room with her. Some cats, especially Oriental types, may panic during labour, especially if it is their first. They may abandon their kittens and follow their owner around crying. These cats may need to be sedated, so it is best to contact your vet.

Kittens are born at varying intervals, most within 30 to 60 minutes of each other. Occasionally several hours may lapse between kittens. If the cat is not distressed and is not straining, you need not be concerned. If she seems weak or distressed and is not interested in the kittens she already has produced, call the vet. If she is straining for longer than 20 minutes and no kittens are produced, seek veterinary assistance immediately.

Cats are wonderful mothers and rapidly clean the kittens and eat the placenta (afterbirth). Many cats purr blissfully throughout the whole procedure. If for any reason your cat doesn't clean the kittens at the time of birth, you will need to intervene. Use clean hands to clear the membranes from around the kitten's mouth and a rough towel to firmly rub its body. Holding the kitten head-down helps to allow fluid to drain out of its airways.

Rearing

The kittens should feed naturally within 10 to 20 minutes of birth. If any seem to be having difficulty then have them checked by your veterinarian. Allow your cat unlimited access to food and water while she is feeding the kittens.

Kittens can start on solid food at two to three weeks old. Provide commercial kitten food for them. You might also offer commercial kitten milk. Do not offer cow's or goat's milk because these contain the milk sugar lactose which the kittens may not be able to digest properly.

TOP: The birth process – giving birth, eating the afterbirth and suckling the kittens.

BREEDING AND REPRODUCTION
Rearing kittens

Provide two litter trays and change them regularly during the day. Kittens will naturally start using litter trays from three to four weeks.

Your nursing cat should be wormed monthly and the kittens should be wormed every two weeks until they are three months old, starting at two or three weeks. Flea treatment of the mother should also be continued during feeding, using a product that is non-toxic to the kittens. The kittens will not need separate treatment until they are weaned.

Kittens may be weaned from seven to eight weeks old. Your cat will be spending less time with them by this stage and they will be taking lots of solid food.

It is important that kittens receive human contact between the ages of two to seven weeks. If they do not they will be shy or aggressive towards people and it will not be as easy to find them a home. The more exposure they have to novel, non-threatening stimuli, including other animals and noises, the more easily they will settle into their new homes.

A newborn kitten cannot see or hear, although it does have a strong sense of smell.

After about a week, the kitten's eyes open and its hearing begins to develop.

At three weeks, the kitten begins to find its feet.

The six-week-old kitten is well on its way to becoming self-sufficient.

The development of a kitten's senses

THE MATURE CAT
A happy retirement

During the last couple of decades the average life expectancy of pet cats has increased by at least two years, largely the result of better nutrition and health care. There are now more elderly cats in the feline population, and like elderly humans, they need special care.

Signs of old age

With increasing age comes a gradual deterioration in health, and while nothing can be done to stop the aging processes, it is possible to minimize their effects. So take note of the signs of old age, and give your cat the extra care it deserves.

Changes to coat and claws

Most cats don't develop many grey hairs as they get old, but the aging process tends to cause their coat to grow longer, even in shorthair breeds. As their joints become less mobile they are unable to groom themselves properly, so their formerly smooth and sleek coat begins to look unkempt, with the fur 'broken' open. Their claws grow more quickly so nail trimming is required more frequently.

Deep sleep

Another sign of old age is increasingly deeper and more prolonged sleep. Old cats are more likely to be startled if woken suddenly, and some cats may even hiss or spit if you wake them unexpectedly by touching or stroking them.
- Allow your elderly cat to sleep peacefully in places of its choice, where it can relax in comfort.
- Warn children in the household not to disturb it.
- Keep other pets away as much as possible.

Changes in feeding and drinking patterns

Your older cat may experience a loss of appetite or a reluctance to eat – it could have difficulty in eating or drinking, too. These symptoms are commonly associated with gum inflammation (gingivitis), tartar formation or tooth decay, which are common problems in elderly cats.

An older cat may also experience increasing thirst. This may be a sign of developing kidney disease or some other health problem.

Old cats may benefit from an adjustment to their diet so that it is more easily digested. Many vets recommend a diet that contains lower protein levels, to lessen any stress on the kidneys. Not all vets agree, so talk to your practitioner, who may recommend a special therapeutic diet.

It is a good idea to take your cat for frequent health checks and routine blood samples to monitor its kidney and liver function.

TOP: Older cats need extra care. Grooming is sometimes difficult for them and they may appreciate regular combing.

Weight loss

This can occur over several months, even though the cat is still eating well, but because the weight loss is gradual, you might not notice it. One cause may be an overactive thyroid gland, a condition that can be treated.

Digestive problems

As a cat ages, it may experience tooth or gum problems that make it difficult to chew, and its digestive system may not function as efficiently. Symptoms of digestive problems include vomiting of food or bile-stained (yellow-green) saliva, diarrhoea and constipation. These may respond to:

- feeding three to four small meals daily (as for kittens)
- including a greater proportion of moist foods in the diet
- feeding products containing more easily digested forms of protein (e.g. lightly boiled egg yolk)
- changing to a prescription diet on the advice of your vet.

Arthritic changes and osteoarthritis

An early sign of arthritis and osteoarthritis is stiffness when getting up and first moving around; this stiffness improves as the day progresses. In more extreme cases the cat has difficulty walking, with weakness of the hind legs, lameness and symptoms of pain.

Two-thirds of osteoarthritis cases in cats are first recognized by their owners, so as soon as you notice any signs you should talk with your vet and follow his or her advice.

Treatments include:

- a non-steroidal, anti-inflammatory drug, given daily in tablet form for long-term treatment
- drugs that aid the production of joint fluid
- various homeopathic and natural remedies, including green-lipped mussel extract, herbal remedies and shark cartilage.

ABOVE: Coat colour changes can occur as part of the aging process (just as people turn grey), though these changes can also indicate poor health.

Reduced bladder capacity and loss of bladder control

A very old or arthritic cat has a reduced bladder capacity. One of the earliest signs of this may be frequent visits outside or to a litter tray. In the later stages it may begin to lose control over its bladder (urinary incontinence), and small amounts of urine may leak onto the places where it sits, lies or sleeps.

Constipation

An old cat has more difficulty in passing faeces, and these may be passed at irregular intervals. Arthritic changes may also prevent it from adopting a normal excretory posture. Because cats usually defecate outside, you may not notice these symptoms for some time, so as your cat begins to age you should keep a closer watch on it.

If constipation develops, talk to your vet about possible corrective measures, which may include the addition of small amounts of medicinal paraffin to the cat's food or a change to a prescription diet.

Increasing deafness

In the early stages deafness may be difficult to detect, because it occurs gradually and many cats learn to adapt. One of the first signs may be failure to respond to your usual calls.

As your cat's hearing deteriorates it will be more prone to accidents. It may not hear vehicles that drive onto your property or approach it on the road.

Increasing blindness

During its early stages deteriorating sight may not be detected. Watch out for the following:

- The eyeballs appear a bluish colour (the cornea is affected)
- The eye appears white in the centre (a sign of cataract)
- The cat starts to blunder into objects such as furniture
- The cat is reluctant to go out at night and/or in bright sunlight.

The same principles apply as for humans. Avoid moving furniture, and protect the cat from danger. A partially or completely blind cat can usually live a contented life as long as it is in familiar surroundings.

ABOVE: Some cats cope better with old age than others. This cat has survived well into its 20s, with an excellent quality of life.

THE MATURE CAT
Caring for an elderly cat

Senility

Signs are:
- Disorientation
- Restlessness
- Increased demand for your attention
- Increased vocalization.

Just like elderly people, old cats have good days and bad days. As the 'owner' you will have to adapt, and be tolerant and sympathetic to the cat's needs. As the situation progresses you will probably need an increasing amount of advice and involvement from your veterinarian, and medication may be required.

Caring for an elderly cat

You can reduce stress on your aging cat by following these suggestions:

- Put comfortable blankets or rugs in its favourite lying places, out of full sun and away from damp areas.
- Protect it from situations in which it is likely to fall: for example, put barriers across steps or stairs, and make sure that it cannot fall from a sundeck.
- If it shows less interest in eating, then slightly warm the food or change the diet to something more palatable.
- Adjust its food intake according to its level of activity. As a cat takes less exercise it tends to put on weight, and an overweight cat is more prone to heart disease. Ask your veterinarian about diets that are specially formulated for various health conditions such as aging kidneys.
- Monitor how much water it is drinking. If the quantity appears to be increasing, talk to your vet.
- Take your cat for regular health checks at your vet clinic. Booster vaccinations need to be kept up to date, and teeth and gums checked. Routine blood sampling may assist with health care.
- If you have to go away, arrange for a housesitter or an alternative home environment rather than a cattery.

ABOVE: Older cats appreciate extra warmth and comfort, as they tend to spend a large part of their day asleep.

79

Thinking about a replacement

As your cat begins to age, you may decide to bring another cat or a kitten into the household. You will need to spend a little time integrating the two animals and working at preventing inter-cat aggression, but it creates a transitional stage that may help you cope with the impending loss of an old friend while adapting to the different demands of a kitten or young cat.

Alternatively, you may decide to wait. Nursing an old cat may be enough for you to cope with, without having to take time away from this older companion to spend time with a younger one.

If you are in any doubt about what to do, talk to your veterinarian and/or the veterinary nurses at the clinic. They will have plenty of experience of owners who have been through the same difficult situation as you, and should be able to offer sound advice.

Whenever you get a replacement, you will have to decide what sort of cat to get, and will probably need to learn a new set of skills for a new individual.

The final days

This can be the most difficult period of your whole relationship with your cat, yet in many ways it can be one of the most rewarding times, too. This is your final opportunity to repay the companionship that your cat has given to you during the happy times you have spent together. If you know what to expect during these remaining days, you will find the inner strength to cope with them and know that your care and concern are being noted.

As your cat becomes more frail, its reliance on you will increase, and nursing procedures will take up more of your time.

As its sense of smell deteriorates, your cat will be less able to detect the aroma of foods, and will therefore become more fussy about what it eats. You will need to talk to your vet about the problem, and try out different types of food to discover what the cat prefers.

Loss of bladder and bowel control may result in 'accidents' to clean up, and if the cat sleeps in a basket its bedding may need frequent changing and washing.

Increasing deafness and blindness will make life more difficult for both of you, and your cat may become disoriented and demand more attention from you day and night.

Give your cat that attention. Physical contact, and the message of love that it carries, is very important; so spend as much time as necessary gently stroking and cuddling your cat to let it know that you are there, and how much you care. Sometimes an old or dying cat will purr a lot more. Why this happens is not fully explained, but for the owner it can be a comforting sound during what is often a stressful time.

Making the decision

Sometimes the final decision is made for you, and the cat dies suddenly and naturally.

In most cases there is no such solution and you, the owner, will have to make the decision to authorize euthanasia. It may come easily, or you may find it very difficult. We witness death (human and animal) on television, videos and films many hundreds of times every year, yet most of us have been brought up in a society that does not cope particularly well with dying, and are unprepared for it in real life.

If there are children in your family, discuss the situation with them and allow them to express their emotions. Talk about the positive aspects that came from owning the cat, and explain that however good their health care may be, cats have a much shorter life expectancy than we do.

The overriding factor in your decision must be to do what is best for the cat, not for yourself or your family. That decision will probably be made with the help of your veterinarian, who can play an important role as a counsellor and advisor.

Veterinarians and their staff understand what you are going through. They deal with this situation almost every day, and many of them have been through a similar situation with a cat of their own. They understand your grief and sense of loss, but also know that they can end your cat's suffering in a humane way.

THE MATURE CAT
The final days

ABOVE: Animals have long been favourite subjects for artists, and a portrait of your pet is a wonderful memento of the times you shared together.

The grieving process

Grieving is a natural human reaction to the death of a much-loved cat or other pet, and you need to express it. There are five well-documented stages in the grieving process, and you will pass through each of them to some degree or another.

1. *Denial and depression.* Confronted with the fact that your cat is at the end of its life, you will probably suffer from depression to a greater or lesser extent. It is often at the subconscious level, and not immediately apparent to those around you. You may tell yourself, 'They've got it wrong', 'Things may not be as bad as they seem', or 'There must be something else that can be done'. This reaction cushions your mind against the emotional blow it is experiencing.

ABOVE: It is important to involve all the family members in decisions about euthanasia, as everyone needs time to say goodbye.

THE MATURE CAT
The grieving process

2. *Bargaining.* In the human grieving process this involves offering some personal sacrifice if the loved one is spared. It is less likely to happen when a pet is involved, but you may still say things to yourself such as, 'If you get better I promise to let you sleep on my bed'.

3. *Pain and anger.* Your emotional pain and feeling of frustration evokes anger. This may be directed at somebody else, such as a close relative or even your veterinarian, or it may be directed at yourself, and emerge as a feeling of guilt. At this stage the support of your veterinarian may be particularly helpful, because negative feelings are not constructive and need to be replaced by positive thoughts.

4. *Grief.* By this stage the feelings of anger and guilt have gone. Your cat has died, and all that remains is a feeling of emptiness. The less support you get at this stage, the longer that feeling will last. If support doesn't come from your family or friends, then get it from another source, such as your veterinarian, a pet cemetarian or a professional counsellor.

5. *Acceptance and resolution.* It will take time, on average three to four months, but eventually your grieving will come to an end. Fond memories will replace grief, and appreciation will replace the sense of loss. Your deep feelings for your cat will remain, but now they will be positive as you recall the happy times the two of you spent together. You may even celebrate them by obtaining a new pet.

Knowing the stages of the grieving process, and how your family, friends and the staff at your veterinary clinic can help, should enable you to come through the event with the minimum of pain and the maximum of love.

Our pet animals have a shorter life span than we do, and on average an owner will suffer the loss of a dearly loved pet five times or more during a lifetime. Each time such a loss occurs that owner will experience grief. It doesn't get any easier the more often it happens, for each pet is an individual and the owner grieves individually for it.

Euthanasia

The usual procedure for euthanasia is the injection of an overdose of anaesthetic into a vein. There is no pain, and the cat falls asleep within 15–20 seconds. You may wish to be present during this procedure, or you may prefer not to witness it and say your last farewell afterwards. The choice will be yours.

Your veterinarian and the veterinary nurses who have assisted will understand what you are going through, and your tears are a natural reaction. One of their obligations is to help you deal with your grief.

Burial or cremation

Your veterinarian can help you to decide what to do next and, if necessary, find somebody who can help to make arrangements. You may wish to have your cat cremated, in which case a casket or urn containing the ashes can be returned to you. This can be buried, or kept. If you wish your cat to be buried, it can be done in your garden or in a pet cemetery.

Counselling

For some pet owners the lingering grief can become intolerable. If this happens to you, don't continue to suffer – rather seek help. This may come from a traditional counselling service, but in some countries (particularly the USA) there are veterinary teaching institutions with social workers specially trained to counsel pet owners. Once again, your veterinarian should be able to advise you.

ABOVE: The joints of an older cat become less flexible, and it is not able to reach every part of its body with its tongue for grooming. Its once glossy coat therefore begins to look 'broken open'.

PROTECTING YOUR CAT'S HEALTH
The veterinary clinic

To keep fit and well your cat needs regular health care. Some of it will be provided by you, and some by your local veterinary clinic.

Veterinary clinics are not just centres for the treatment of ill health. They are also a valuable source of practical information, specialized products and friendly advice from veterinarians and their trained nursing staff.

Most veterinary clinics also act as valuable community resource centres, providing information about local boarding facilities, cat groomers or cat-sitting services, for example. Many also have notice boards on which their clients can post information, such as photographs of missing pets or kittens looking for a new home.

The changes in veterinary science over the years, and particularly during the last decade, have been remarkable. In addition to radiography (X-rays) and routine blood sampling, modern diagnostic aids include ultrasonography (the use of ultrasound scanning equipment), Magnetic Resonance Imaging (MRI) and Computer Assisted Tomography (CAT) scans.

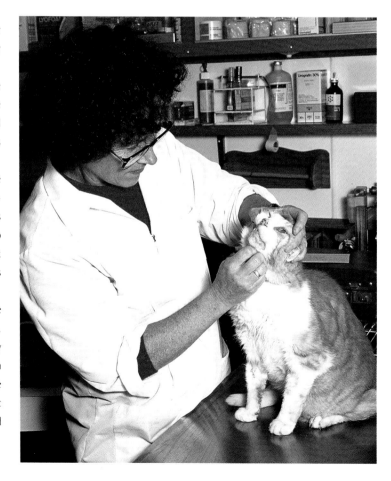

ABOVE: Once your cat is 10 years old or so, it should get a thorough veterinary check-up every year.
TOP: Radiography is only one of the many diagnostic procedures available to veterinary surgeons.

Other areas of veterinary specialization include:
- Anaesthesia
- Orthopaedics
- Ophthalmology
- Endocrinology
- Dermatology
- Animal Behaviour
- Dentistry
- Medicine
- Surgery
- Radiology
- Diagnostic imaging

Certain methods of diagnosis and therapy – some ancient, some new – are also now becoming a recognized part of a comprehensive approach to animal health care. Known as Complementary Veterinary Medicine (or Complementary and Alternative Medicine – CAVM), many of these methods have been used in human medicine for years, but their integration into veterinary practice has been comparatively recent. For example, some veterinarians are now trained in veterinary acupuncture and acutherapy (the examination and stimulation of specific points using acupuncture needles, injections, low-level lasers, magnets and a variety of other techniques for diagnosis and treatment); veterinary chiropractic (the examination, diagnosis and treatment through manipulation and adjustments of specific joints, particularly the vertebrae, and of areas of the skull); veterinary massage therapy; homeopathy; botanical medicine, nutritional therapy and even the use of flower essences (dilute extracts from certain flowers).

TOP: Chiropractic treatment is one of many alternatives to traditional veterinary medicine.

The immune system

The bodies of animals, like those of humans, have various defence mechanisms to protect them against microorganisms in the environment.

Healthy skin acts as a physical barrier, while the mucous membranes in the nose, trachea and bronchi help to trap foreign substances and prevent them from entering the lungs. Other primary barriers include acid in the stomach, which kills many invading organisms, and mucus produced from the lining of the intestines. The liver destroys toxins produced by bacteria.

These defence mechanisms work well when an animal is healthy, but are less effective if it is weakened, unhealthy, or mentally or physically stressed.

Most organisms that cause disease consist mainly of proteins. If an organism gets past the primary barriers, the body quickly detects its 'foreign' proteins and produces antibodies against them. Antibodies are produced by specialized white blood cells found mainly in the lymph nodes and spleen. They circulate in the blood and are usually very specific, destroying only the organism (the antigen) that stimulated their production.

The first time the cat's body encounters a specific disease, introduced from the environment or by means of a vaccine, it may take up to 10 days to produce antibodies. The next time the disease is encountered, antibody production occurs very rapidly, preventing the disease from becoming established.

Antibody levels wane with time, but if the antigen is encountered again (either through infection or a booster vaccination), antibody production immediately resumes. Immunity created by vaccines is not generally as long lasting

TOP: Kittens receive antibodies in their mother's milk, which protect them against certain diseases until they are between six and 12 weeks old.

PROTECTING YOUR CAT'S HEALTH
The immune system

as the 'natural immunity' created by exposure to a disease, which explains why booster vaccinations may be needed to keep an animal protected.

Passive (maternal) immunity

Passive immunity occurs when a newborn animal acquires antibodies from its mother.

Newborn animals have a rudimentary immune system that takes many weeks to fully develop. To tide them over this period they receive passive immunity from their mother in the form of antibodies, some of which enter their body while they are still in the uterus, but most of which are taken in with the mother's colostrum, or first milk. This is a critical period for the newborn animal, which can only absorb these antibodies during the first day or two after birth. If a queen has a prolonged kittening or gives birth to a large litter, the early kittens will have more opportunity to ingest colostrum than those born later, so the degree of passive immunity may vary between littermates.

A queen can only pass on antibodies to the diseases that she herself has encountered, or against which she has been vaccinated. Therefore a queen that is not vaccinated, or lives in isolation from other cats, will have fewer antibodies to pass on, and her kittens will be more vulnerable from birth. A queen used for breeding must be vaccinated, and her booster vaccinations kept up to date.

Passive (maternal) immunity is only temporary. Passive immunity wanes over time, the amount of antibodies in the blood halving every seven days or so. In most kittens the level of maternal immunity will have fallen almost to zero by the age of 12 weeks.

Active immunity

Active immunity is the result of an animal producing antibodies from its own immune system, in response to disease or vaccination.

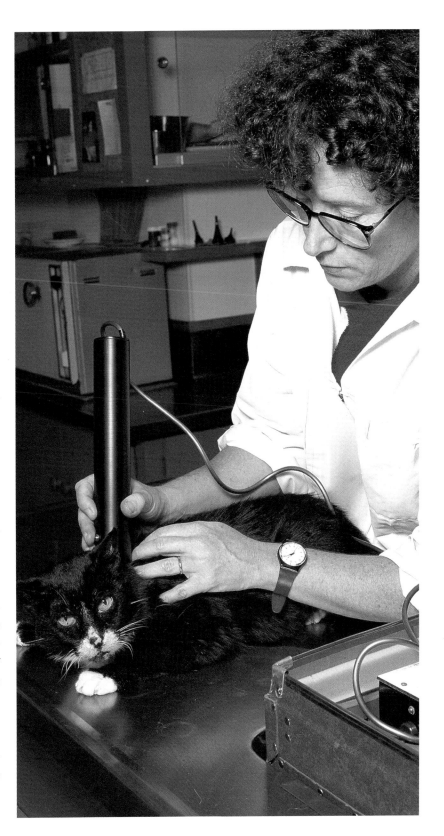

ABOVE: Painless laser therapy can be effective in relieving muscular aches and pains and can speed up the healing of ligament injuries.

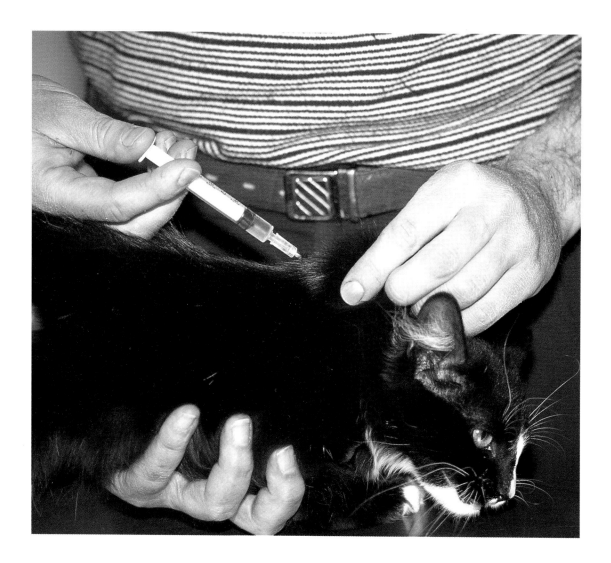

To be protected, kittens must develop their own, active immunity, either by contact with a specific disease or through vaccination.

While its passive immunity is high the kitten is protected from disease, and its own immune system may not respond to a vaccination, although some brands of vaccine are designed to override it and stimulate the kitten's immune system.

Although we know that the level of passive immunity steadily decreases, we cannot be sure at what age each individual kitten will entirely lose this immunity and respond to vaccination. In some kittens this can occur much earlier than 12 weeks, so these will be at risk if they are not vaccinated and are exposed to a virus.

The typical recommendation for a kitten from a queen that has been properly vaccinated is to receive two vaccinations, the first at eight or nine weeks of age and the second (booster) four weeks later. In areas of high risk, vaccination may be started at six weeks of age and repeated at fortnightly intervals until 12 weeks. Your vet will advise you if this is necessary.

Vaccinating your cat

In many countries routine vaccination programmes have greatly reduced the incidence of several important feline diseases. Many different brands of vaccine are available, including multiple vaccines that are effective against several

TOP: Soon after you get your new cat, register it with your local vet, who will keep up-to-date records of vaccinations and booster dates, saving you from having to remember these details yourself.

PROTECTING YOUR CAT'S HEALTH
Feline respiratory diseases

diseases. Your veterinarian will advise you which vaccines are the most suitable for your cat.

Feline respiratory diseases

Many different organisms can cause respiratory infection in cats, but two in particular are equally responsible for about 90 per cent of cases, and for the disease complex that is commonly called feline influenza or cat 'flu.

Feline herpesvirus-I is a virus similar to the one that causes cold sores in humans. The disease that it causes is called feline viral rhinotracheitis (FVR), and is highly contagious. Initial signs are sneezing, fever, and a discharge from the eyes and nose that quickly becomes purulent due to secondary infection by bacteria. As the disease progresses an affected cat may develop ulcers in the mouth, bronchitis and eventually pneumonia. A pregnant queen may abort her kittens. Although not many adult cats die from the disease, the death rate amongst kittens can be 50–60 per cent. Recovered cats often carry the virus for years. Much of the time they are not infectious to other cats, but every now and again they go through episodes during which the virus is shed.

A similar percentage of feline respiratory infections are caused by feline calicivirus (FCV). In these cases ulceration in the mouth, nose, and on the tongue, is common. Other symptoms are similar to those of FVR, although the disease may be less severe. Recovered cats become carriers of the virus, which may then be persistently shed.

Respiratory infections due to a combination of both the above viruses are not uncommon.

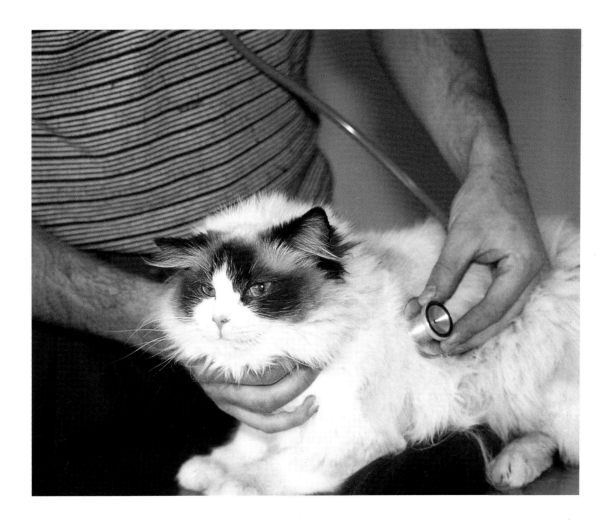

ABOVE: Heart murmurs are common in cats, and a heart check should be part of your cat's annual check-up.

Feline infectious enteritis (FIE)

Also known as feline panleucopaenia (FPL), this was once one of the most common, widespread and serious diseases of domestic cats. As a result of vaccination programmes it is now well under control. It causes a dramatic drop in the numbers of circulating white blood cells, and symptoms include fever, loss of appetite, vomiting, depression, and diarrhoea. Kittens are most susceptible and there is a high mortality rate. Cats that survive often remain debilitated for the rest of their lives.

Feline leukaemia virus (FeLV)

This is the most important cause of feline cancer. Symptoms are variable, and can include vomiting, diarrhoea, lethargy and laboured breathing.

Various types of treatment may be tried, including chemotherapy, but these are often unsuccessful.

On average FeLV affects about one to two per cent of cats, but the disease is more common in some countries than others so you should talk to your veterinarian about the incidence in your area, and whether your cat needs to be vaccinated.

About five per cent of respiratory infections are caused by *Chlamydia psittaci*, an organism called a 'rickettsia' that is classified as intermediate between a virus and a bacterium. It causes a disease that was once called feline pneumonitis and, unlike the viruses mentioned above, will respond to certain antibiotics. It causes runny eyes (conjunctivitis) and nose (rhinitis), although fever or more severe respiratory symptoms are uncommon, and deaths are fortunately rare. It can be particularly troublesome where there are groups of cats, such as in boarding catteries or breeding establishments.

Rabies

In countries where this viral disease is endemic, cats are routinely protected by vaccination. Rabies can affect any mammal and is almost always fatal. Transmitted through the saliva of infected animals, usually the result of a bite, it can also be spread by infected saliva coming into contact with mucous membranes (eye, nose or mouth) or a skin wound.

In Europe foxes are the most important carriers, while in North America raccoons, bats, skunks, foxes and coyotes are the culprits. In Mexico and other Latin and Central American countries, cats are the most common carriers.

TOP: A not-too-unusual friendship: introduce a kitten to this little family pet and they could become great companions – do remember, though, that guinea pigs and cats can transmit fleas to one another.

PROTECTING YOUR CAT'S HEALTH
Rabies and its prevention

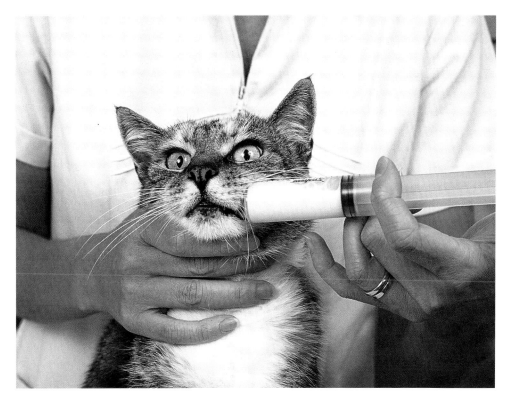

The incubation period is usually two to eight weeks, but can be up to six months. The virus travels via the nerves to the brain where it causes inflammation (encephalitis) resulting in nervous symptoms. In the last stages of the disease the virus moves into the salivary glands and saliva.

In its early stages rabies commonly causes changes in behaviour and personality. Affected animals become anxious and increasingly sensitive to noise and light. Nocturnal animals may be seen out during the day, and wild animals may lose their fear of humans. A normally timid cat may become more friendly, while a normally friendly cat may become shy and hide away from light.

As the disease progresses, affected cats may become restless and irritable, and they are likely to attack other animals or humans without provocation. They eventually develop paralysis of the throat and cheek muscles, which makes it impossible for them to swallow – as a result, saliva drools from their mouth. Breathing becomes increasingly difficult, and in the final stages the animal collapses, enters a coma and dies.

Prevention

In certain countries where rabies is endemic their law requires the vaccination of cats and dogs. Many island nations, in which the disease is not endemic, have strict quarantine laws to prevent its introduction. In Britain a scheme has been introduced that allows vaccinated cats and dogs entry under certain conditions (see PETS Travel Scheme, pp42–3).

If your cat fights with any mammal that is a rabies carrier, saliva carrying the virus could be present on that cat's coat or in any of the wounds inflicted on it.

If you think your cat has been in a fight with a rabid animal:
- Don't try to capture the attacking animal.
- Take extreme care when handling your pet. Use gloves, and cover it with a towel.
- Allow as few people as possible to handle it.
- Call Animal Control or an equivalent organization.
- Take your cat to a veterinarian.
- If your cat has a current rabies vaccination, get advice about giving a booster within 72 hours (this is compulsory in the USA).
- In the USA, if your cat does not receive a booster within 72 hours, then unless the attacking animal tests negative your cat will be quarantined for six months at a veterinary clinic or disposed of by Animal Control.

If you are bitten or scratched by an animal you suspect is rabid, or if its saliva enters an open wound or comes into contact with your nose, eyes or mouth, wash the wound or contact area using household detergent or soap. These kill the virus faster than any disinfectant. Get immediate medical attention – treatment involves a course of vaccinations.

ABOVE: The easiest way in which to administer liquid medicine is with a syringe.

Take these routine precautions to prevent rabies:
- Don't feed or attract wildlife into your yard.
- Call Animal Control if you suspect there is a rabid animal in your yard. Don't try to capture wildlife.
- Don't allow bats to live in your attic or chimney.
- Don't pick up dead or abandoned animals.
- If you are particularly at risk (eg if you regularly handle dead animals or nervous tissue), ask your physician if you should be vaccinated.

External parasites

External parasites live on or in the skin of the cat. Most external parasites are host-specific, which means that they infect one particular species of animal only – among the exceptions is the cat flea, which may also infect dogs.

Fleas

Most cats are likely to become infected with fleas at least once during their life. Sources of infection include other cats, dogs, hedgehogs and even rabbits. The flea most commonly found on cats is the cat flea (*Ctenocephalides felis*).

The major natural factor controlling the flea population is not so much the temperature but the humidity, because the cat flea cannot effect development if the humidity is less than 50 per cent. In winter, even in homes that are centrally heated, the humidity is around 40 per cent, so fleas are less of a problem. In summer humidity and air temperature both rise, so the flea population increases.

The most common symptom of flea infestation is the cat repeatedly scratching, and nibbling or licking at its fur – some cats may even become skittish and edgy, as if they are trying to run away from their fleas. Fleas occur in greater numbers on some areas of the body than others; especially along the back just in front of the tail. You can see if there are any fleas present by grooming the cat with a flea comb, which will comb out either the live fleas or their droppings (flea dirt). To identify the latter, squeeze them between two pieces of damp tissue. They contain digested blood, which will stain the tissue reddish-brown.

Cat fleas remain on a cat only long enough to feed and lay eggs. The latter quickly fall off into the environment, and contaminate the cat's bedding and household carpets. Flea control must include treatment of the cat (and any other pets), and also the household environment.

There are various treatment preparations available, including liquids, powders, sprays, collars (which may cause a skin reaction) and medallions that are worn by the cat and gradually release an insecticide. Environmental treatments include 'flea bombs'. Talk to your vet about what products are best for your particular area.

TOP: Flea powders have now been largely superseded by easy-to-use topical liquid preparations.

PROTECTING YOUR CAT'S HEALTH
External and internal parasites

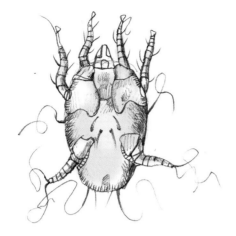

Ticks

These are more common in rural areas, and often attach to a cat's head or neck. To remove them, swab them with alcohol or methylated spirits for a short time, grip them as close to the skin as possible with a pair of tweezers, then pull them off.

Some ticks (particularly in Australia) are toxic and can kill small animals such as cats. Your local vet will have up-to-date information.

Mites

The ear mange mite (*Otodectes cynotis*) results in irritation that causes the cat to scratch at its ears. In doing so it often introduces a secondary bacterial infection, and the ear becomes inflamed and painful. If your cat often scratches at its ears, get a vet to check them. Treatment usually involves ear drops or an ointment. Use a product prescribed by your vet.

A very small mange mite (*Notoedres cati*) may burrow into the skin, especially around the sides of the face and between the eyes and ears. It causes irritation and skin thickening, and scratching and licking by the cat results in hair loss and baldness. This type of infection can take some time to treat, and veterinary advice is essential.

The larvae of the harvest mite (*Trombicula autumnalis*), variously called red bugs or chiggers, can infect cats during the summer and early autumn. They are usually found on less furry areas of a cat's skin, such as the ears, sides of the mouth, and between the toes. Various insecticides are available, so talk to your veterinary clinic.

Fur mites (*Cheyletiella* species) are common on cats, dogs and rabbits – each type of animal has a specific mite species. These mites may cause itching, but the most common sign is profuse dandruff, especially along the cat's back and sides. Although not particularly serious to cats, this mite can also infect humans, on whom it causes itchy weals and blisters that progress to dry scaling. Areas most commonly infected are the hands and forearms, and sometimes the chest.

The sarcoptic and demodectic mange mites are not uncommon on dogs, but are rarely seen on cats.

Lice

Healthy cats are unlikely to become infected, but debilitated or sick cats may contract an infection because they are unable to groom themselves properly. The lice (*Felicola* species) lay white eggs (nits) that are firmly attached to the hairs. Various insecticides are available for the treatment of this condition.

Internal parasites

Regular veterinary checks (including, if necessary, examination of faecal samples) and regular de-worming should ensure that your cat does not suffer from these internal parasites. Ask your vet clinic for advice about the treatment procedures best suited to your cat's particular infestation.

Roundworms

The ascarid worms, *Toxocara cati* and *Toxascaris leonina*, are the most common roundworms in cats. They can grow up to 10 cm (4 in) long, and lay eggs that under suitable conditions can remain viable in the environment for years. Infection may be direct (from eggs) or indirect (from eggs that have hatched into infective larvae in an intermediate host such as a mouse or rat).

In most parts of the world, about one cat in every five is infected with *Toxocara cati*. While infected adult cats may show few, if any, signs of ill health, the larvae of this worm can infect suckling kittens via the mother's milk, and kittens can be seriously affected.

ABOVE: When infected by the roundworm *Toxocara cati*, adult cats show few signs of ill health.
TOP: The most common mange mite to infect the cat is the ear mange mite, *Otodectes cynotis*.

Hookworms

These tend to be a problem in warmer and more humid areas, so are more common in Australia, New Zealand, parts of the United States, and South Africa than in the United Kingdom. The *Ancylostoma* species is the most important, and a heavy infestation can cause severe anaemia and even death. Veterinary treatment is essential.

Whipworms (*Trichuris*) and threadworms (*Strongyloides*)

These mainly occur in the warmer, more humid areas such as parts of the United States and Australia, and are rarer in cats than in dogs. The life cycle is direct.

Tapeworms

The most common tapeworms in cats are the *Dipylidium caninum*, transmitted by a cat eating infected fleas or lice during grooming, and *Taenia taeniaeformis*, transmitted by a cat eating infected prey such as rats or mice.

ABOVE: Tapeworm segments may be shed in a cat's faeces, or adhere to the fur at the rear end.
TOP: Getting their own back: mice infected with tapeworm (*Taenia taeniaeformis*) will transmit the parasite to their killer, the cat.

PROTECTING YOUR CAT'S HEALTH
Internal parasites

Flukes

Areas of risk are mainly North America and tropical regions. Most infections are the result of cats eating raw fish. Veterinary treatment is possible.

Lungworms

A number of different species of lungworm can infect cats, but most cause few problems.

Aleurostrongylus abstrusus needs veterinary treatment, though, and symptoms range from mild coughing to severe breathing problems.

Heartworms

Although more common in dogs, these parasites can infect cats, especially in warmer areas such as around the Mediterranean, Australia and parts of the United States. Signs include coughing and breathlessness, and sudden death may occur. Treatment is essential.

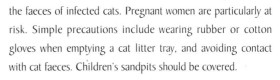

Toxoplasmosis

Infection by the protozoan *Toxoplasma gondii* rarely causes symptoms in cats, but the significance of this parasite lies in the fact that the disease can be transmitted to humans through the faeces of infected cats. Pregnant women are particularly at risk. Simple precautions include wearing rubber or cotton gloves when emptying a cat litter tray, and avoiding contact with cat faeces. Children's sandpits should be covered.

Feline infectious anaemia (FIA)

The incidence of this disease varies in different parts of the world. It is caused by *Haemobartonella felis* (or *Eperythrozoon felis*), a protozoan parasite transmitted by biting insects such as mosquitoes. The parasites destroy the red blood cells, and early symptoms include pale mucous membranes and lethargy. The disease is usually noticed by the cat's owner during its early stages, and can be treated with antibiotics. However, it is often associated with feline leukaemia virus (FeLV) and in these cases the chances of recovery are poor.

ABOVE: The fungal infection ringworm causes bald patches on the cat's skin, and is highly contagious to humans.
TOP: Avoid direct contact with cat faeces, as it can transmit a number of parasites, such as the protozoan *Toxoplasma gondii* – wear rubber gloves as a precaution.

MONITORING YOUR CAT'S HEALTH
Signs of ill health

The earlier you can detect a health problem and do something about it, the better. Treatment is more likely to be effective, and your cat will probably suffer less discomfort or pain. Learn what is normal, so that you can detect when something abnormal occurs.

Early signs of ill health

One of the first signs of ill health may be a subtle change in your cat's normal behaviour. It may be quieter than usual, less active, or disinclined to go for a walk. It may be more thirsty, or less hungry. Since cats, like humans, have their 'off' days, you should keep an eye on this sort of change for a day or two. If it continues, then take further action.

Consult your veterinarian if your cat shows any of the following signs:

- unusual tiredness or lethargy
- abnormal discharges from the nose, eyes, ears or other body openings
- excessive head shaking
- excessive scratching, licking or biting at any part of the body
- markedly increased or decreased appetite
- excessive water consumption
- difficult, abnormal or uncontrolled waste elimination
- marked weight loss or weight gain
- abnormal behaviour such as hyperactivity, aggression or lethargy
- abnormal swellings on any part of the body
- lameness
- difficulty in getting up or down.

As soon as you see anything unusual, make a note of it, for you may need this information if you need to take your cat to a veterinarian at a later date. Doctors talk to their human patients to get a 'history' of their problem before making an examination and reaching a diagnosis. Vets cannot ask their animal patients about the problem, and rely on their human owners for such a 'history'. It will be helpful if you can supply as much information as possible.

Pain

Pain results from the stimulation of specialized nerve endings (receptors) in the body. It has many causes, but is usually the result of injury, infection, poisoning or inflammatory reaction. It is one of the earliest signs of disease.

TOP: Kittens should be vaccinated against infectious diseases at eight or nine weeks old, and as adults they will need annual boosters.

MONITORING YOUR CAT'S HEALTH
Spinal, head and internal pain

When we suffer pain, we can tell somebody. A cat cannot speak, but in most cases your cat's reactions will be fairly clear to you.

- It will usually cry in pain if you accidentally step on its foot or if something strikes it. It may cry out if you touch a painful part of its body, and even hiss, bite or lash out at you.
- If it has hurt a leg it may put its foot on the ground but place no weight on it, limp, or carry the leg off the ground.
- If it is suffering from joint pain, for example from arthritis, it may cry out when getting up or lying down.
- Pain or irritation from its anal glands will cause it to 'scoot' its bottom along the ground. In response to hind-end pain it may frequently inspect the affected area.
- Pain in an eye will cause it to paw at the affected area or rub it against objects.
- Pain in an ear usually results in it tilting its head to the affected side and shaking it.
- Mouth pain may cause it to salivate and judder its jaws.

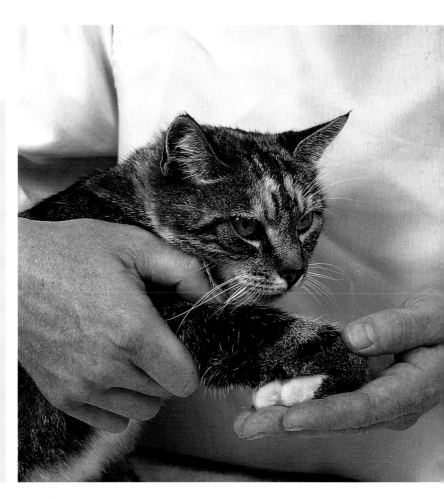

It can be more difficult to detect when a cat is suffering spinal, head or internal pain. The only indication that something is wrong may be a less obvious change in its behaviour.

Suspect spinal pain if your cat:
- seems to be lame, but no limb is affected
- resents being touched along its back
- humps its back and/or trembles when standing up
- is incontinent
- has difficulty assuming the normal posture for defecation
- collapses on its hindquarters.

Suspect head pain (headache) if your cat:
- has half-closed eyes, but no obvious eye problem
- presses the top of its head against objects
- gently but regularly shakes it head
- stares vacantly.

Suspect internal pain if your cat:
- spends more time lying down
- is particularly restless, and unable to settle down
- keeps its abdominal muscles tensed or stands in a hunched-up position
- continually strains to pass a bowel movement, but fails to do so
- is usually docile and inexplicably becomes aggressive.

What to do
If the pain was caused by a minor accident (e.g. somebody stepped on the cat's toe), use common sense and monitor the outcome. If the pain persists after a few hours, you should contact your vet.

If pain is the result of something more serious, or if you are not able to determine the cause, you need to seek advice from your veterinary surgeon.

TOP: There is not always an obvious reason for lameness. Your veterinarian is the best person to determine the cause.

Blood and circulatory system problems

SIGNS	SOME POSSIBLE CAUSES	ACTION
Exercise intolerance, lethargy, weakness, fainting (kitten or young cat)	Congenital malformation in which blood bypasses the lungs	Take to the vet in each case
As above, poor growth, distended abdomen	Congenital defect in which blood bypasses the liver	
As above (any age)	Heart valve defect	
As above (any age)	Anaemia	
Coughing	Congestive heart failure (chronic heart disease) Heart-based tumour Heartworm disease (see p95)	Take to the vet in each case
Abnormal breathing	Lung congestion due to poor heart function Warfarin poisoning Anaemia	ALL URGENT. Take to the vet
Pale or bluish tinge to gums	Warfarin poisoning Poor heart function Abnormal destruction of red blood cells Blood clotting disorder	Take to the vet in each case
Jaundice (yellow tinge to gums and whites of eyes)	Abnormal destruction of red blood cells Secondary to liver infection	Take to the vet in each case
Abdominal distension	Fluid accumulation due to poor heart function	Take to the vet
Loss of use of hind limbs, howling, hind feet feel cold	Aortic thrombosis (blood clot blocking one or both femoral arteries)	URGENT. Only about half of cats recover with treatment

MONITORING YOUR CAT'S HEALTH

Blood and circulatory system — ear problems

Ear problems

SIGNS	POSSIBLE CAUSES	ACTION
Shaking head, scratching ears, blackish discharge	Ear mites	Take to the vet
Shaking head, reddish or yellowish-white smelly discharge. Ear painful to touch	Inflammation of ear canal and inside of ear flap. External ear infection (*otitis externa*), usually caused by a mixture of bacteria, fungi and yeasts. Can cause severe pain and discomfort and lead to permanent ear damage	Do not place anything into the ear, as drum may be ruptured and some treatments may be harmful if they enter the middle ear. Vet will check that ear drum (tympanic membrane) is intact, and may take a swab to determine which infective organisms are present. Follow the treatment prescribed by your vet
Head tilted to one side, loss of balance, abnormal movement of eyeballs (*nystagmus*). Cat may be shaking head	Middle and inner ear disease (*otitis media* and *otitis interna*). A foreign body penetrating the ear drum, or chronic ear infection	Take to the vet. Treatment may involve anti-inflammatory drugs, antibiotics and drugs to stop any vomiting
Scabs on white ears	Sunburn (white cats are particularly susceptible)	If the scabs are superficial and skin is reddened, it is probably sunburn. Apply paediatric sunblock three times daily. Keep cat out of intense sunlight
	Cancer of the ear (squamous cell carcinoma). White cats are particularly susceptible	If the scabs are deep, long-standing and have never healed, it may be skin cancer. Take to the vet. Treatment may involve freezing with liquid nitrogen or surgical removal of the pinna

Endocrine problems

SIGNS	SOME POSSIBLE CAUSES	ACTION
Abdominal enlargement, excessive thirst, symmetrical hair loss, pigment changes	Hyperadrenocorticism or Cushing's syndrome (excessive production of adrenalin)	Take to the vet
Excessive thirst, increased appetite and urination, weight loss (middle-aged or old cat)	Diabetes mellitus	Take to the vet
Neck swollen, hyperactive, increased appetite, vocalization and urine output, excessive thirst, weight loss	Hyperthyroidism (excess of thyroid hormone)	Take to the vet

Eye problems

SIGNS	SOME POSSIBLE CAUSES	ACTION
Avoiding light, blinking	Several	Take to the vet
Runny eyes, clear discharge	Wind, dust, strong sunlight, allergy or blocked tear duct. May have no tear duct	Bathe with cooled, boiled water or ophthalmic saline. If it doesn't clear, take to the vet
Runny eyes, clear or purulent discharge, whites of eyes inflamed, pawing at eyes	Bacterial or viral conjunctivitis	Take to the vet
As above, one eye only	Possible foreign body in eye, or injury	Take to the vet
Tacky, purulent discharge, eye surface dry, conjunctiva inflamed	Dry eye (*keratoconjunctivitis sicca*)	Take to the vet
Eyes appear to be white, cat's vision affected	Cataract formation	Take to the vet
Cat appears blind, no other symptoms	Retinal degeneration	Take to the vet
Closing one eye, avoiding light, pain, eye watering, blood in the eye	Inflammation within the eye (*uveitis*)	Take to the vet

MONITORING YOUR CAT'S HEALTH

Endocrine – eye – intestinal problems

Eye problems *(continued)*

SIGNS	SOME POSSIBLE CAUSES	ACTION
White line or dot on eye surface, pain, eye watering	Corneal ulcer, often result of a cat scratch	Take to the vet
Third eyelid showing	Nerve damage	Take to the vet
Pressing head against objects, eye protruding, avoids light	Glaucoma (swelling of the eyeball due to accumulation of fluid)	Take to the vet

Intestinal problems

SIGNS	SOME POSSIBLE CAUSES	ACTION
Eating well but thin	Worm burden	Treat for worms
Vomiting and/or diarrhoea (may be intermittent), weight loss	Inflammatory bowel disease (IBD) – the inflammation of the bowel and a reduction in its ability to absorb nutrients. May be bacterial overgrowth. Because affected cats are not obtaining enough protein, they may lose weight	Take to the vet
Flatulence	Usually dietary. More common in kittens	Talk to vet about changing diet
Chronic weight loss despite normal or increased appetite	Worms, intestinal tumour, inability to absorb nutrients	Talk to vet to confirm cause
Vomiting, not eating	Inflammatory bowel disease (see above)	Take to the vet
Hunched-up posture	Foreign body Severe constipation Abdominal pain	Take to the vet in all cases
Straining to defecate, hard firm faeces, straining ceases after passage of faeces, not vomiting	Mild constipation, particularly common in elderly cats and longhaired cats	Give one teaspoon medicinal paraffin. If faeces are not passed within eight hours, take to the vet
Straining to defecate, few/no faeces, depressed, possibly vomiting	Severe constipation	Take to the vet to ascertain cause

Intestinal problems

SIGNS	SOME POSSIBLE CAUSES	ACTION
Diarrhoea, one or two bouts, without blood, cat otherwise bright and alert, no vomiting	Food intolerance Mild bacterial enteritis	Water only for one day, then bland diet for 24 hours. If diarrhoea stops, gradually re-introduce diet. If continues, see vet
As above, after cat has drunk cow's milk	Lactose intolerance	Feed commercial low-lactose milk
Diarrhoea, frequent and persistent, cat otherwise bright	Giardiasis (*Giardia* infection) Coccidiosis	Take to the vet in each case
Diarrhoea, frequent, may be blood, depressed, abdominal pain	Bacterial enteritis (e.g. *Salmonella* or *Campylobacter*) Colitis, tumour	Take to the vet in each case

Liver, spleen and pancreatic problems

SIGNS	SOME POSSIBLE CAUSES	ACTION
Abdominal distension, with or without jaundice	Liver tumour	Take to the vet
Vomiting, jaundice, dark-coloured urine, abdominal pain, poor appetite	Bile duct blockage Bile duct rupture	Take to the vet in each case
Vomiting, diarrhoea, jaundice	Feline infectious peritonitis (FIP)	Take to the vet
Acute, persistent vomiting, fever, abdominal pain	Pancreatitis. Pancreatic digestive enzymes are secreted into the pancreatic tissue itself, causing inflammation and tissue destruction. Can cause death. Recovered animals may suffer permanent dysfunction of the gland	Take to the vet
Excessive thirst, hunger, may be abdominal enlargement, lethargy, weight loss	Diabetes mellitus. If the pancreas does not produce enough insulin, glucose levels build up in the bloodstream. Glucose then passes through the kidneys into the urine, taking water with it	Take to the vet for blood and urine tests. Some cases may be controlled by adjusting the diet; most cases require insulin injections given at home

MONITORING YOUR CAT'S HEALTH

Intestinal – liver, spleen or pancreatic – mouth

and oesophagus problems

Liver, spleen and pancreatic problems (*continued*)

SIGNS	SOME POSSIBLE CAUSES	ACTION
Cat wobbly and uncoordinated, may progress to depression. Cat 'spaced-out'. May bump into things. If not treated, will collapse	Hypoglycaemia (low blood sugar) due to too much insulin or too little food	Hypoglycaemic coma may occur in a cat receiving insulin injections if it is subject to excessive activity or exercise, goes an excessive length of time between feeds, or has missed a feed
		Inadequate glucose levels in the bloodstream are further lowered by the action of the injected insulin, leading to collapse, coma and convulsions. Treatment is to give glucose or honey by mouth. Owners of diabetic cats should always keep these substances on hand in case of an emergency
Very depressed and prostrate. The breath may smell like nail polish remover (acetone)	Ketoacidosis (build-up of ketones in the bloodstream) may occur if the diabetes is not properly controlled and blood sugar reaches excessively high levels	Take to the vet
Cat comatose	May be either ketoacidosis or hypoglycaemia	URGENT ACTION. Do not attempt to treat. Take to the vet

Mouth and oesophagus problems

SIGNS	SOME POSSIBLE CAUSES	ACTION
Bad breath	Tartar build-up on teeth	Take to the vet. Teeth may be scaled and polished under anaesthetic
Bad breath, bleeding gums, difficulty in eating	Gingivitis (inflammation of gums)	As above
Difficulty in eating, bad breath, jaw juddering, dribbling	Broken or infected tooth	Take to the vet for tooth extraction. Other teeth may need attention

Mouth and oesophagus problems (*continued*)

SIGNS	SOME POSSIBLE CAUSES	ACTION
Drooling, pawing at mouth, may be gulping	Foreign body (e.g. bone or stick) lodged across the hard palate, or fish hook in lip. Cut tongue (due to fighting or licking out of cat food tin)	If possible, open mouth and check. Remove foreign body if possible. Otherwise, take to the vet
	Bee sting in mouth (on tongue, inside cheeks or gum)	If possible, open mouth and check. Remove sting with tweezers. Check mouth regularly, and if more than slight swelling take to the vet
	Ulcerated tongue	If tongue inflamed or ulcerated, investigate access to irritant poisons. Take sample of suspected substance to the vet
Drooling, retching or coughing	Object stuck in throat Tumour in mouth	As above Take to the vet in each case

Nervous system problems

SIGNS	SOME POSSIBLE CAUSES	ACTION
Loss of balance, uncoordinated	Middle ear infection Vestibular disease (infection, inflammation or tumour affecting the vestibule) Brain tumour Disease of cerebellum	Take to the vet in each case
As above, cat's diet contains raw fish	Thiamine deficiency	Take to the vet for thiamine injections, and change the diet
Muscle spasms, fits or convulsions	Epilepsy Poisoning Brain tumour	Take to the vet
Muscle spasms, fits or convulsions, late pregnancy or within eight weeks of giving birth	Eclampsia (lowered calcium levels in bloodstream)	URGENT ACTION. Get vet treatment with calcium injection
As above, may be head pressing, pain around head	Encephalitis or meningitis	URGENT ACTION NEEDED Take to the vet
Collapsing, third eyelids visible, limbs rigid, tail straight, facial muscles contracting	Tetanus infection	Take to the vet

MONITORING YOUR CAT'S HEALTH

Mouth and oesophagus — nervous system — reproductive — respiratory system problems

Nervous system problems (*continued*)

SIGNS	SOME POSSIBLE CAUSES	ACTION
Salivating, may be other signs	Poisoning	Take to the vet in each case
Salivating, behavioural change	Rabies	
Abnormal head position, eyes may be flicking from side to side	Middle ear disease Vestibular disease (infection, inflammation or tumour affecting the vestibule) Brain tumour	Take to the vet in each case
Collapsing in hindquarters, with or without acute pain	Disc protrusion in thoracic or lumbar region	Take to the vet
Collapse, walking in circles or partial paralysis, eyelids partly closed, eyes flickering	Stroke	Take to the vet

Reproductive problems in the queen

SIGNS	SOME POSSIBLE CAUSES	ACTION
Persistent oestrus	Ovarian cyst	Take to the vet
Excessive thirst, reduced appetite, vomiting, distended abdomen, vulval discharge, 6–8 weeks after end of oestrous cycle	Pyometra (accumulation of pus in uterus)	URGENT ACTION. Take to vet. Surgical removal of uterus and ovaries may be necessary
Enlarged mammary gland, not painful or painful and inflamed	Mammary tumour (not always malignant) Mastitis	Take to the vet in each case
Lethargy, appetite loss, within 1–2 weeks of kittening. May be purulent discharge from the vulva	Metritis (inflammation of uterus)	Take to the vet. May need antibiotics or spaying

Respiratory system problems

SIGNS	SOME POSSIBLE CAUSES	ACTION
Sneezing, clear nasal discharge	Viral infection Allergy (e.g. to pollen) Blade of grass lodged in nose Feline viral rhinotracheitis (FVR) Feline calicivirus infection (FCV)	If persistent, take to the vet in each case

Respiratory system problems (*continued*)

SIGNS	SOME POSSIBLE CAUSES	ACTION
Sneezing, purulent nasal discharge from one or both nostrils	Tumour Bacterial or fungal infection Molar abscess	Take to the vet in each case
Red, puffy skin on nose, with some crusting	Allergy (e.g. to mosquito bites) Sunburn Early skin cancer	If red and sore, keep out of the sun. Apply paediatric sunblock. If does not subside, take to the vet
Noisy breathing	Laryngeal problem (e.g. laryngitis) Allergic bronchitis	Take to the vet
Rapid breathing	Pneumonia Heart problem Allergic asthma Poisoning (e.g. aspirin) Ruptured diaphragm after fall or accident Pyothorax (pus in chest, usually the result of a cat bite)	Take to the vet URGENT. Take to the vet in each case
As above, gums pale or white	Internal or external haemorrhage Poisoning (e.g. warfarin)	URGENT ACTION. Take to the vet in each case
Choking, collapse	Foreign body in throat obstructing breathing	Try to remove the obstruction Take to vet urgently
Abdominal movement associated with breathing	Ruptured diaphragm after accident/trauma Pneumothorax (air in the chest, usually after accident/trauma) Pyothorax (see above) Haemothorax (blood in chest) after poisoning with anticoagulants (e.g. rat poison) Rib/lung damage from trauma/accident	Take to the vet in each case
Bleeding from nose	Acute trauma Foreign body in nose Problem with clotting mechanism Poisoning with rodenticide (e.g. warfarin) Tumour	Take to the vet in each case

MONITORING YOUR CAT'S HEALTH
Respiratory – skeletal, joint and muscle problems

Respiratory system problems (*continued*)

SIGNS	SOME POSSIBLE CAUSES	ACTION
Coughing, mild, occasional	Tracheitis Allergy Heart problem	Take to the vet in each case
Coughing, frequent, shallow, history of accident	Pneumothorax (air in chest)	Take to the vet
Coughing, frequent, harsh, associated purulent nasal discharge, cat appears ill	Feline viral rhinotracheitis (FVR)	Take to the vet

Skeletal, joint and muscle problems

SIGNS	SOME POSSIBLE CAUSES	ACTION
Slight lameness on one limb, one joint mildly painful when flexed or extended	Sprain (slight damage to ligament or cartilage in a joint)	Get veterinary advice
Sudden lameness, bleeding foot	Cut pad	Take to the vet
Sudden lameness on one hind leg. Touching toe to ground but not bearing weight on it	Ruptured anterior cruciate ligament in knee joint. Usually the result of an accident, e.g. if caught in a fence or the branches of a tree, twisting the leg and snapping the ligament	Take to the vet. This may need surgery
Sudden lameness, holding one hind leg off the ground	Slipped knee cap (patella)	Take to the vet
Sudden lameness, hind limb painful	Hip dislocation Fracture of head of femur	Take to the vet in each case
Sudden lameness after fall or accident, swelling over a section of the affected limb, pain	Bone fracture	Take to the vet
Sudden lameness, swelling of tissue on leg	Bite wound	Take to the vet. Wound may form an abscess
Sudden hind limb collapse, cat usually in pain	Fractured pelvis	Take to the vet
Difficulty getting up or down, eases out of stiffness after activity	Arthritis (degenerative joint disease)	Take to the vet

Skeletal, joint and muscle problems (*continued*)

SIGNS	SOME POSSIBLE CAUSES	ACTION
Difficulty in assuming normal posture for urination/defecation	Spondylosis (degenerative condition resulting in deposition of extra bone between vertebrae)	Take to the vet
Firm, painful swelling just above a joint, becoming larger over period of time	Osteomyelitis (bone infection)	Take to the vet
Chronic lameness in one leg	Arthritis (degenerative joint disease)	Take to the vet

Skin problems

SIGNS	SOME POSSIBLE CAUSES	ACTION
Scurfy skin, white flakes in coat	*Cheyletiella* (mite) infection	Ask vet for insecticidal treatment
Scurfy, itchy skin on head and shoulders. See tiny insects	Lice (see p93)	As above
Hair loss, symmetrical, no irritation, no broken hairs	Hormonal imbalance	Take to the vet
Broken hairs and asymmetrical loss	*Psychogenic alopecia*, self-inflicted by overgrooming in response to stress	Take to the vet
Hair loss and scaly skin. Not itchy	Ringworm (fungal infection)	Take to the vet
Scratching, skin red, appears wet, on back of neck or inner thighs	*Eosinophilic granuloma* complex	Take to the vet
Scratching, excessive licking, may be skin change	Allergy to fleas, food or environment. Mites (e.g. *Demodex* or *Notoedres*)	If flea control is adequate, take to vet. Or treat for fleas (see p92)
Scratching, chewing, pussy and inflamed skin, may be bleeding	Pyoderma (deep bacterial infection)	Take to the vet
Ulcer on lip or nose	*Eosinophilic granuloma* complex	Take to the vet
Lump or swelling within the skin. Not painful	Lipoma (fatty tumour) Haematoma (blood blister) Skin tumour Sebaceous cyst	Take to the vet in each case
Lump or swelling within the skin, painful, may be discharging fluid	Abscess	Take to the vet

MONITORING YOUR CAT'S HEALTH

Skeletal, joint and muscle — skin — stomach —

urinary problems

Stomach problems

SIGNS	SOME POSSIBLE CAUSES	ACTION
Eating grass, then vomiting grass and mucus. Hair or prey remains may be included in vomitus	Natural evacuation of indigestible material	Carry out protocol for vomiting (see p121)
As above, fluid only	Mild gastritis	Carry out protocol for vomiting
Vomiting, longhaired cat	Fur ball	Talk to vet about diet and laxatives or oil. Comb more often
Vomiting, frequently, refusing food, depressed.	Gastritis Pancreatitis Obstruction (could be a fur ball)	Take to the vet in each case
As above, plus diarrhoea (with or without blood), dark tarry faeces	Feline infectious enteritis (FIE) Feline leukaemia virus (FeLV) Poisoning	Take to the vet in each case
As above, plus hunched posture	Foreign body lodged in stomach Pancreatitis	Take to the vet, urgent action required
Distended abdomen, young cat, may be lethargic, poor coat	Worm burden	Treat for worms (see pp93–5)

Urinary problems

SIGNS	SOME POSSIBLE CAUSES	ACTION
Excessive thirst, bad breath, large quantities of urine, mouth ulcers, weight loss, anaemia, vomiting	Chronic kidney disease (nephritis) resulting from an infection, chronic degeneration, tumour or inherited defect	Measure cat's water intake over a day. Take this info to the vet
Young animal, failing to thrive, excessive thirst, very pale urine	Juvenile renal disease (inherited)	Take to the vet
Smelly urine, may contain blood, frequent urination or urine leaking, may be licking vulva or penis	Cystitis (inflammation of bladder) due to infection, bladder stones or stress	Take to the vet
Male cat, straining to pass urine, may be vomiting and yowling	Urethral blockage, possibly by a bladder stone	URGENT ACTION REQUIRED Take to the vet immediately

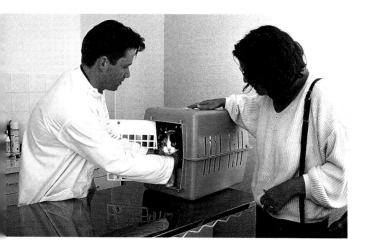

FIRST AID
Emergency treatment

The information given below is for guidance only, and is not intended to replace veterinary advice. If faced with an emergency, remember that the principles of first aid for cats are similar to those for humans.

A basic first-aid kit
- Rolls of 2.5 cm (1 in) and 5 cm (2 in) bandage
- Self-adhesive bandages
- A 2.5 cm (1 in) crepe bandage
- Roll of 5 cm (2 in) by 7 cm (3 in) adhesive plaster
- Non-stick gauze pads
- Cotton wool
- Tweezers
- Curved, blunt-ended scissors
- Straight scissors
- Nail clippers
- Antiseptic and disinfectant liquids
- Tube of antiseptic cream
- Hydrogen peroxide (3 per cent) for flushing wounds
- Medicinal paraffin for treatment of constipation
- Small nuggets of washing soda to induce vomiting
- Ear and eye drops as recommended by your veterinarian
- Roll of absorbent paper towel

TOP: It is sensible to take your cat to the vet in a box or carrier. This helps it to feel secure and removes the risk of the cat escaping from your arms when frightened by a dog in the waiting room, or the sound of a passing car on the street.

FIRST AID

Restraining a cat and administering tablets

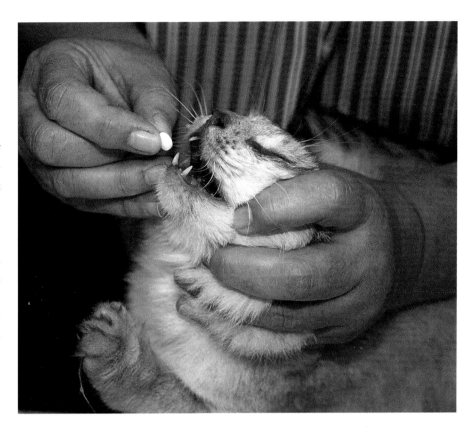

Administering tablets to a cat

Place the thumb and fingers of one hand on each side of your cat's mouth. Press in gently and tilt the cat's head back – its mouth should open.

Use your other hand to pull down on the lower jaw, then place the tablet as far back on the tongue as possible.

Hold the cat's mouth closed and head back until the tablet is swallowed.

Restraining a nervous cat

When you need to take a nervous or distressed cat to the veterinary clinic, it may be best to restrain it in a blanket. Your vet will then be able to unwrap the calmed cat to examine it.

Administering a liquid to a cat

Tip the cat's head back. Pull out the corner of the lip to make a pocket, then pour the liquid into it. Hold the cat's mouth closed until the liquid is swallowed – you'll find this difficult to achieve, though, and may need to adjust the dosage accordingly, to take spillage into account (be careful not to overdose).

Minor wounds

A cat's natural instinct is to lick and clean up any wounds it suffers. A wound exposed to the air will usually dry up and heal more quickly.

Where minor wounds are involved you can usually clip off the surrounding hair, then check for and remove thorns, glass or other embedded objects. Flush the wound with saline or three per cent hydrogen peroxide, then leave the cat to look after itself.

Do keep an eye on any minor wound, though, because if the licking becomes excessive the cat may cause skin changes and introduce infection – if this happens, ask your veterinarian for advice.

It is often difficult for an owner to bandage an affected area, and in many cases a cat will remove a bandage soon after it has been applied. Adhesive tape can be applied over a bandage to reduce this risk, but this should be done by a veterinary nurse or a veterinarian. Your vet may recommend an Elizabethan collar as a last resort to prevent a cat from licking at a wound.

If the wound is on an area that the cat cannot reach, clip off the hair and flush the wound with saline solution (two tablespoons salt in two cups water), three per cent hydrogen peroxide or an antiseptic solution recommended by your vet.

Bite wounds and puncture wounds

Some of these may appear minor, but by their nature they have the potential to cause problems. The opening will heal over very quickly, and any infection that has been introduced (which is often the case) will be trapped inside and may form an abscess.

The cat may lick the puncture wound and keep it open, or you can do the same by frequent bathing with saline. If you are in any doubt about what to do, or if the wound appears infected, get veterinary advice, because antibiotic treatment is usually necessary.

FIRST AID
Dealing with wounds

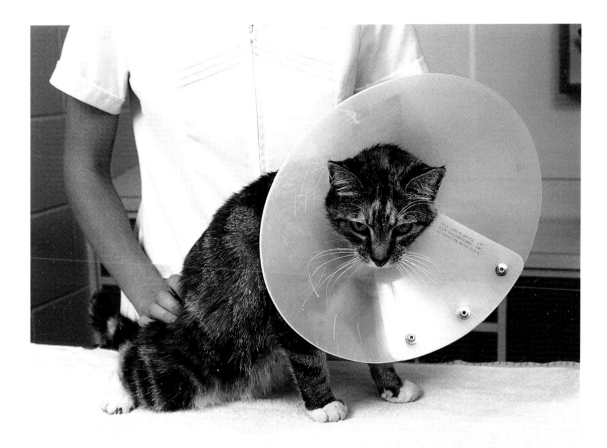

Bleeding from a vein

If your cat is bleeding from a vein, the blood will be seeping out and dark in colour. Try to flush the wound with saline or three per cent strength of hydrogen peroxide.

If the wound is on a limb, try to apply a pressure bandage as follows:

- Cover the wound with a non-stick gauze pad.
- Place a thick pad of damp cotton wool over the top.
- Bandage firmly (but not tightly). Use good-quality bandages and keep the tightness even.
- Check at regular intervals to make sure that there is no swelling below the wound (a sign that the bandage is too tight).
- You could bandage the limb all the way to the foot and envelop the foot, to prevent this type of swelling.
- Arrange for a veterinary check.

If the wound is on an area that you cannot bandage, apply the non-stick pad and cotton wool, then use your thumb or fingers to apply gentle pressure for up to five minutes at a time. If bleeding continues, get help as soon as you can.

Bleeding from an artery

Arterial blood is bright red and spurts vigorously.

- If the blood vessel or artery involved is not too large, apply a pressure bandage as above and check every 10 minutes to make sure that bleeding has stopped.
- For larger blood vessels and arteries, use your fingers to apply firm pressure over the affected area, but slightly closer to the heart. Release pressure after five minutes, then re-apply if necessary.
- As soon as possible, take the cat to a vet.

TOP: Elizabethan collars may be used to prevent cats from licking a wound or lesion on the body, or damaging surgical sites on the ears or eyes. Cats should not be allowed to go outside while wearing these collars.

Bleeding from a nail

This is often associated with an accident, and in most cases the nail will have been torn completely off. The wound will usually be painful and the cat probably will not allow you to touch it.

If you are able to do so, protect the wound by placing a sterile pad over the injured area, then wrap a bandage around the whole paw. Blood clotting should occur within five minutes, but the wound will need further treatment and antibiotic cover, so contact your veterinary clinic.

Eye injuries

You can deal with minor problems, such as dust or dirt, by flushing the eye with the eye drops kept in your first-aid kit, or an eye solution for humans.

Remember, though, that the surface of the eye (cornea) is fragile and damage to it may not become visible for several days. For this reason, keep a close watch on any eye problem and get it checked if you are in any doubt.

Ear injuries

These may result from a cat fight, or the cat catching on an obstruction such as a twig. Surface veins on the ear are easily damaged and bleeding may occur. Unless the wound is minor, get a veterinary check.

Mouth injuries

These are usually caused by a sharp bone. Minor wounds to tongue or gums will usually heal without incident, but if in any doubt get a veterinary check-up.

To remove a fish hook caught in a cat's lip, try to push the barb all the way through, then cut the hook through its shank. If you can't do this yourself, get veterinary help.

Leg and paw injuries

Wounds on the lower leg can be flushed with saline or three per cent hydrogen peroxide, then dressed, bandaged and protected by an old sock. Cuts on a footpad are more difficult to treat, and unless they are minor, they are best left for a veterinarian to assess.

Tail injuries

Wounds may be caused during fighting, so treat as described above (Leg and paw injuries).

If there is severe pain around the affected area, the tail may be fractured, so get veterinary advice. Sometimes the tail gets trapped under a vehicle tyre, and in trying to run away the cat pulls the tail violently, causing severe damage to the nerves. In many cases the tail will be paralysed – unfortunately the damage will be permanent and the only treatment is tail amputation.

TOP: Care should be taken when applying bandages. If they are too tight, they may restrict the blood flow. Cats can be difficult to bandage when conscious, especially if in pain – this cat is under sedation.

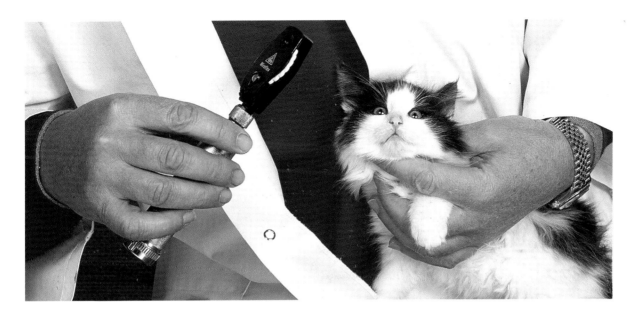

Fractures

If a leg is fractured, restrict the cat's movement as much as possible. You can provide support to an injured lower limb by tying a newspaper or magazine around it. If in doubt, leave well alone or you could aggravate the existing damage. Get professional help.

For first aid to fractured ribs, use any materials you can find to wrap around the whole chest. Take the cat to a vet.

Accidents and emergencies

Before you begin to administer any emergency treatment, you will need to check for the cat's heartbeat. To do this, place two fingers over the lower centre of the chest just behind the elbow of the front leg, and press down lightly.

Remember not to move the cat unless it is in danger (for example, if it is in the road).

Artificial respiration (AR)

- Remove the cat's collar and wipe away any saliva, blood or vomit. Pull the tongue forward.
- Place the cat on its side.
- Place one hand over the cat's mouth to keep it closed.
- Take a deep breath and blow strongly into the cat's nostrils for about three seconds until you feel resistance or see the chest rise.
- Repeat this procedure 12–15 times over one minute.
- Stop, and watch the chest to see if the cat is breathing on its own.
- If the cat is not breathing, continue to perform the above procedure.
- Get veterinary help as soon as possible.

Cardiopulmonary resuscitation (CPR)

As with AR, this can be difficult to administer because of a cat's small size.

- Place the cat on its right side.
- Spread the fingers and the palm of one hand over the cat's chest.
- Apply smooth, rhythmical compressions that will move the chest about 2 cm (0.7 in) but not cause internal injury. Press once a second for half a minute.
- Stop, check for heartbeat (see left, Accidents and emergencies).
- If there is no response, repeat the CPR procedure for another half a minute, then perform artificial respiration (see left) for about one minute.
- Stop, and check for heartbeat and breathing.
- If neither is present, continue.
- If heartbeat only is present, continue AR.
- Get veterinary help as soon as possible.

TOP: This vet is using an ophthalmoscope to check a kitten's eyes for injuries.

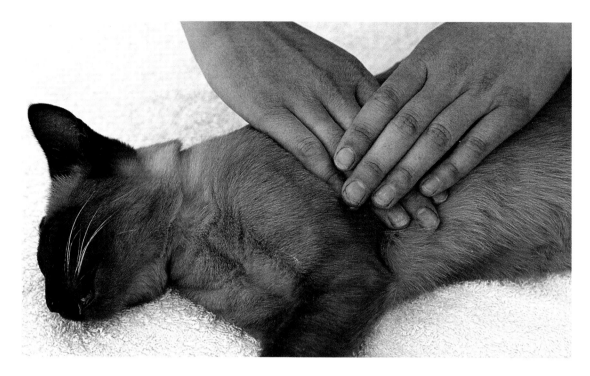

Road traffic accidents

- If the cat is on a roadway, get somebody to warn or control approaching traffic.
- Carefully slide the cat onto a piece of cloth, clothing or something similar and move it off the road to a safe area.
- Check for heartbeat and breathing.
- If the cat has a heartbeat but is not breathing, perform artificial respiration (see p115).
- If the heart is not beating, perform CPR (see p115).
- An injured cat that is conscious will be frightened, aggressive, or both. Act and move as calmly and quietly as possible, talking to and reassuring the cat all the time.
- Use the cloth or clothing to load the cat into a vehicle for transportation to a veterinary clinic. Excessive handling could exacerbate some injuries, especially spinal fractures.
- If you think there is a spinal injury and you have the materials on hand, slide the cat onto a solid board and tie it down to restrict its movement.
- If the only way you can move the cat is to pick it up, do so very carefully using one hand in front of the chest and the other under its rump, keeping its spine as straight and as still as possible.

Shock

Most accident victims suffer shock. Signs of shock are:
- rapid breathing
- pale or white gums
- rapid heartbeat.

What to do

- If the cat is unconscious, lay it on its side and pull its tongue out to keep its airway open, then place something under the cat's hips to raise its hindquarters.
- If the cat is conscious, try to calm it and gently restrain it.
- Try to stop any visible bleeding (see p113).
- Keep the cat warm, but do not apply heat.
- Get veterinary help as soon as possible.

Electrical shock

- If the affected cat is still in contact with the electrical source, SWITCH OFF THE POWER before you touch the cat.
- Check for heartbeat and breathing.
- If there is a heartbeat but no breathing, try AR.
- If the heart is not beating, try to perform CPR.
- Get the cat to a vet as soon as possible.

TOP: Do not attempt CPR unless it is impossible to get to a vet, as you may cause severe damage by performing this procedure incorrectly.

FIRST AID
Shock, drowning and burns

Drowning

- If you can, hold the cat upside down by its back legs and swing it from side to side for 15–20 seconds to help water to drain out of its lungs.
- Lay the cat on its side, sloping with its head down.
- Check for a heartbeat and breathing.
- If there is a heartbeat but no breathing, try to perform artificial respiration (see p115).
- If there is no heartbeat, try to perform CPR (see p115).
- Get the cat to a vet as soon as possible.

Carbon monoxide, smoke or other vapour inhalation

- Carry the cat into fresh air.
- If the cat is conscious, flush out the eyes with clean water.
- If the cat is unconscious, check its heartbeat and whether it is still breathing.
- If the heart is beating but there is no breathing, try to perform artificial respiration (see p115).
- If the cat has no heartbeat, try to carry out CPR (see p115).
- Get the cat to a vet as soon as possible.

Choking

- If possible, get help to restrain the cat.
- Use the fingers and thumb of one hand to press the upper lips over the teeth in the upper jaw. Further firm pressure will force the mouth open.
- If you can see the object that is causing choking, try to remove it, but be careful that you don't get bitten.
- If removal this way is not possible and the cat is quite small, hold it by its hind legs, head down, and shake it vigorously.
- Get the cat to a vet as soon as possible.

Convulsions or fits

These usually last for only a few minutes, and are rarely fatal. Your objectives are to prevent the cat from injuring itself while convulsing, and avoid injury yourself.

- Keep your fingers away from the cat's mouth.
- Move it to a clear area away from furniture.
- Wrap the cat in a blanket to help restrain its leg and body movements.
- Contact your veterinarian for advice.

Burns

Burns may be caused by heat, or by chemicals such as petroleum products or strong acids or alkalis.

Burns caused by heat

- Get the cat to a veterinarian as soon as possible.
- While doing so, apply cold water or an ice pack (for example, a pack of frozen vegetables such as peas or corn kernels) to the affected area.
- Try to prevent the cat from licking at the area.
- If the burns are extensive, cover the affected area with a sterile, non-stick dressing.

Burns caused by chemicals

- Thoroughly wash the area affected by the burn with soap (preferably use a soap that is mild, unscented and uncoloured) and water.
- Try to determine the cause.
- Get veterinary advice.

Heat stroke

Signs include rapid, irregular breathing, panting, vomiting and collapse.

- Get the cat into a cool environment.
- If the cat is unconscious, apply artificial respiration and/or CPR as necessary (see p115).
- Use a garden hose or a bath of cold water to cool the cat for up to half an hour. It is also a good idea to place an ice pack (see above) on top of the cat's head.
- Get veterinary help.

Hypothermia

- Warm the cat using an electric blanket or similar heating pad, turning the cat every few minutes.
- Otherwise use a warm (37°C; 100°F) water bottle covered in a cloth.
- Get veterinary help.

Frostbite

The areas most commonly affected on a cat are those with little hair or a minimal blood supply, such as the tips of the ears and the nose.

- Apply a towel or similar material soaked in warm water (24°C; 75°F)
- Check the skin colour. If it appears dark, get veterinary help immediately.

Insect stings and spider bites

A great variety of insects (such as bees, wasps and hornets) and spiders are poisonous. All stings or bites may produce an allergic reaction.

- If the sting is from a bee, use a blunt knife or similar object to scrape off the stinger embedded in the skin. Do not simply try to pull it out.
- Apply ice to the affected area.
- Get veterinary help as soon as possible.

Poisonous toads and lizards

In the United States the Blue-tailed Lizard and at least nine species of toad are capable of poisoning cats. If a cat licks or bites a toad, a toxin (carried in the wart-like lumps on the toad's skin) enters the cat's mouth and (often) its eyes. If it eats the tail of a Blue-tailed Lizard, it ingests the poison contained in it. Clinical signs appear soon after the event, and include salivation, vomiting, shaking or trembling, lack of coordination, convulsions and coma.

- If possible, immediately flush out the cat's mouth and eyes with water.
- If it is unconscious, wrap it in a blanket to keep it warm.
- In all cases take it to your vet for emergency treatment.

Snake bites

Bites may occur from venomous or non-venomous snakes. Venomous snakes leave a distinctly different imprint to that left by a non-venomous species, but these are difficult to see under a cat's fur. Because cats can be difficult to handle, you should always get veterinary help as soon as possible.

If you can handle the cat, try to carry out some or all of the following procedures.

- If you are sure the bite is from a non-venomous snake, clip the hair from the affected area and flush it with three per cent hydrogen peroxide poured onto the bite.
- If you are not sure about the bite's origin, treat it as if it were poisonous.
- For poisonous bites on a limb, make a tourniquet from a belt or a piece of cloth folded to about 2.5 cm (1 in) wide. Place this between the bite and the heart, about 2.5–5 cm (1–2 in) away. Put a small stick or similar object over the tourniquet, tie it with a single knot and twist it just tightly enough to cut off the circulation to the bite area. Wrap a piece of cloth around the stick and limb to keep it in place.
- Wherever the bite area is located, clip the hair from it and use a knife to make a single cut over each fang mark until it bleeds.
- Suck the venom from the area. DO NOT DO THIS IF YOU HAVE ANY OPEN SORES OR CUTS IN OR AROUND YOUR MOUTH.
- Spit out the blood that you suck in. DO NOT SWALLOW IT.
- Flush the wounds with three per cent hydrogen peroxide.
- Apply ice to the affected area.
- Take the cat to your veterinarian.
- If travelling to the clinic takes some time, then every 15 minutes loosen the tourniquet for 10 seconds, then tighten it again.

An encounter with a skunk

In the United States, skunks are one of the major carriers of rabies and should not be handled with bare hands. Follow these procedures if your cat has encountered a skunk and been sprayed over the face or body.

- Restrain the cat.
- Flush its eyes with clean water.
- Wash its body thoroughly with soap and water.
- To neutralize the odour, use a skunk odour neutralizer or liberally apply plain tomato juice.
- If the skunk has died, DO NOT HANDLE IT WITH YOUR BARE HANDS. Take the carcass to a veterinarian for a rabies examination.
- Ensure that your cat is vaccinated against rabies.

FIRST AID
Frostbite and poison

Poisonous substances in and around the home

Safety around the home is just as important for pets as it is for small children.

A liquid or powder that has leaked or been spilt from a container can get onto a cat's fur or paw pads, and in cleaning this off the cat is likely to ingest the toxic substance. You may find that your cat develops a real liking for the taste of bleach (the scented bleaches are more likely to appeal) and will lick it off wiped surfaces.

All potentially poisonous substances should therefore be locked away, or stored safely out of a cat's reach and where a cat or other animal cannot knock them off. Substances that can give off harmful vapours should be used and stored where there is adequate ventilation.

Symptoms of poisoning

Symptoms will vary, depending on what toxic substance is involved, and can be similar to those of many other medical conditions. However, you should consider the possibility of poisoning if your cat is:

- panting very heavily
- suddenly vomiting and/or has severe diarrhoea (more than two or three times within an hour)
- drooling or foaming at the mouth
- crying
- suffering intense abdominal pain
- showing signs of shock
- depressed
- trembling, uncoordinated, staggering or having convulsions
- collapsed or in a coma
- showing signs of an allergic reaction, such as swelling around the face or a red rash (hives) on the belly.

What to do

- Time is critical.
- Try to identify the poison.
- Carry out the recommended emergency treatment described on p120.
- Contact your veterinarian immediately and take your cat to the clinic.
- If you have found the cat with a poisonous or unidentified substance, take the container or packet with you.

TOP: Cats wandering through overgrown garden areas may risk snakebite or encounters with other poisonous creatures.

The label should contain information about the antidote and treatment for that particular type of poisoning.
- If your cat has vomited, collect a specimen of the vomit in a clean container and take that with you.

Emergency treatment

If the poison is CORROSIVE (strong acid or alkali) or petroleum-based, or if you are not sure what caused the problem:
- DO NOT INDUCE VOMITING.
- If the cat is conscious, flush the mouth and muzzle with large quantities of water, then try to give one teaspoon of egg white or olive oil.
- Take the cat to your vet.

If the substance is NOT CORROSIVE (is not strong acid or alkali) or petroleum-based:
- If the cat has not already vomited, induce vomiting.
- Put the vomit material into a clean container.
- Take the cat and vomit material to the veterinarian.

To induce vomiting

Give ONE of the following:
- one large crystal of washing soda straight down the throat
- one heaped teaspoon of table salt in a little warm water
- one tablespoon of mustard powder in a cup of warm water.

Repeat every 10 minutes until the cat vomits. Save the vomit for veterinary examination.

Emergency antidotes

- Absorbents (absorb toxic substances):
  activated charcoal, up to six 300 mg tablets, or two to three tablespoons of powder mixed in a cup of warm water
- Protectants (help to cover the stomach lining):
  one tablespoon of egg white or olive oil
- Against acids:
  one teaspoon of bicarbonate of soda
- Against alkalis:
  several teaspoons of vinegar or lemon juice

Some potential sources of poison

Many of the plants in our gardens, and the substances that we commonly use in the house, garden, garage or shed, can be poisonous to cats and other pets (as well as small children, of course). Kittens are particularly at risk.

Inside your home

Petroleum products
- dry-cleaning solution

Corrosive products
- detergents in concentrated form, such as those for use in dishwashers or automatic washing machines, dry-powder carpet cleaners
- household bleaches (hypochlorites, chlorox)
- disinfectants in concentrated form
- corn and callous remover

Non-corrosive products
- medicines (human or animal). Symptoms include vomiting, panting, acetone odour to breath, general weakness or collapse
- some indoor house plants may be poisonous if eaten, such as poinsettia leaves, mistletoe
- cleaning agents and dry shampoos containing carbon tetrachloride
- chocolate. Contains theobromine, a compound with a similar action to caffeine. It is a stimulant and irritant, and affects every organ of the body. The darker the chocolate, the more theobromine it contains. The amount of theobromine in chocolate intended for human consumption is safe for a human, but can harm a pet animal. Symptoms include digestive upset (diarrhoea and vomiting), increased heartbeat and blood pressure, increased urine production causing excessive thirst, muscle twitching and convulsions. There is no known antidote.

Note: Pet chocolate drops are safe because the theobromine has been removed.

Vapours
- from cleaning agents or solvents such as acetone, benzene, or carbon tetrachloride

FIRST AID

Potential sources of poison

- carbon monoxide from leaking gas appliances, or improperly ventilated oil or solid fuel stoves. It is odourless, colourless and tasteless
- smoke (from cigarettes, cigars or open fires)

In the garage or shed

Petroleum products
- solvents and paint removers
- engine oil

Corrosive substances
- battery acid
- grease remover
- strong alkalis such as lye and other drain cleaners
- creosote and tar

Non-corrosive
- plant and garden sprays and weed killers (fungicides and herbicides)
- insecticides, especially pyrophosphates such as malathion. Potentially lethal. Absorption can occur through the skin
- metaldehyde. Commonly used in slug and snail baits. Although many of these products contain a repellent to reduce the risk to cats, cases of cumulative poisoning do occur. A cat may ingest only a few baits at a time, but over a period enough metaldehyde will accumulate inside its body to cause poisoning. Metaldehyde may also be present in the compressed tablets used to fuel small heaters
- rat and mouse poisons. There are many different products. Active ingredients include arsenic, thallium and warfarin
- antifreeze (*ethylene glycol*). Some cats, especially kittens, like the taste of antifreeze and will lap it if they discover it. It is extremely toxic and causes kidney damage. A tiny amount can be fatal. Symptoms usually begin an hour or two after ingestion.

Vapours
- fumes from wood preserver, or acetone-based paint removers and solvents

In the garden

- toadstools and fungi
- berries such as mistletoe
- plants: azalea, bird of paradise, crocus, delphinium, foxglove, irises, ivy, jasmine, laburnum, larkspur, laurel, lilies, lily-of-the-valley, oleander, privet, rhododendron, sweet pea, wisteria
- certain vegetables: rhubarb leaves (raw or cooked), tomato vines

In the car

- carbon monoxide from a faulty exhaust

In the neighbourhood

Poisonous substances may also be found in public areas away from your home.
- rat poisons, laid by rodent eradicators
- other poisons, such as bird or rabbit carcasses baited with poison to eliminate vermin in woodlands or forests
- food contaminated with Salmonella or *Clostridium botulinum* bacteria. Cats are usually very fussy about what they eat, but *salmonellosis* (food poisoning) can be fatal in young animals. Botulism affects the nervous system and causes partial or complete paralysis.

A PROTOCOL FOR ASSESSING AND TREATING VOMITING

In cats, vomiting is a natural way of eliminating material from the stomach, and just because a cat vomits does not necessarily indicate a problem: it may simply be getting rid of indigestible material such as the remains of a prey victim. If your cat vomits once or twice, and otherwise appears bright, monitor its progress for a few hours. If no more vomiting occurs, offer it small amounts of food over the next 24 hours, and if all is well, re-introduce its normal diet.

Consult your vet if you are in any doubt, or
- the cat appears depressed
- there is blood in the vomit
- the cat is vomiting intermittently (e.g. every three to four hours) for more than eight hours
- the cat is vomiting continuously
- the cat has had access to potentially poisonous substances.

Index

Note: page numbers in italic refer to illustrated material.

A

AAA (animal assisted activities) 18
AAT (animal-assisted therapy) 18, 19
ACA (American Cat Association) 18
accidents and emergencies 115
   carbon monoxide inhalation 117
   cardiopulmonary resuscitation (CPR) 115
   choking 117
   drowning 117
   electrical shock 116
   heat stroke 42, 117
   hypothermia 117
   road traffic accidents 116
   shock 116
   skunk, encounter 118
   smoke inhalation 117
aggression 26–7
   dealing with redirected inter-cat 65
   fear 63
   inter-cat 65
   play 63
   predatory 62
   redirected 61, 64
   status-related 57, 61
   towards cats in the household 64
   towards other cats 64
aging process 76
   coat colour changes 77
anti-anxiety medication 64, 65
antibodies 86–7
artificial respiration (AR) 115

B

balanced diet 51
   meals, commercial 48, 49
bathing 40
bed and bedding 31
behaviourist, animal 65, 67, 71
Birman, breeding cattery 25
birth process 74
bleeding, from a nail 114
   a vein 113
   an artery 113
boarding catteries, standards 37, 43
bones 53
   pressure cooking 51, 53
booster shots 88
breed standards, United States variations 18
breeders 24, 25, 27, 28, 32
breeding 24, 37, 72
breeds 16
   Abyssinian 5, 15, 16, 17
   Angora 14, 15, 16
   Asiatic desert cat 11
   British Sealpoint 17
   British Shorthair 15, 24
   Burmese 14, 16, 38, 57
   Chinchilla 24
   Chinese cat 15
   Forest wild cat 10
   Japanese Bobtail 16
   Korat (Si-Sawat) 24
   LaPerm 24
   Longhair 14, 16, 23, 24, 30, 31
   Maine Coons 16
   Manul (*Felis manul*) 14
   Manx 14
   Norwegian Forest Cat 24
   Ocicats 16
   Oriental 23, 24, 71, 74
   Red Lynx 24
   Persian 14, 15, 24
   Russian Blue 16, 38
   Shorthair 14, 15, 16, 24, 32, 76
   Siamese 14, 15, 16, 21, 38, 57, 59
   Spotted Mist 16
   Turkish Van 24
burial or cremation 83
burns 117
   caused by chemicals 117
   caused by heat 117

C

CAB (Cat Association of Britain) 18
cages 42
   travel 38, 41
cancer (*see* feline cancer)
carbohydrates 46
cardiopulmonary resuscitation (CPR) 115
carrying basket 34
cat doors 34, 67
Cat Fancy Association (CFA) 18, 73
cat-human relationship 57
catnip 64
cats and balance 37
cats, benefits of 19–21
cat-scratch fever 35
cattery 24, 64
cat shows 16, 17, 40
collars 32, 34, 41, 68
communication 32, 33, 58, 60
colour 5, 15, 16
chlamydia psittacci 90
choking 117
claws, regular trimming 71
claw sheath 40
claw-marking 32, 36, 40, 56
coat colours and patterns 14, 15
   range 24
constipation 53
convulsion or fits 117
counselling, for pet owners 83

D

death, preparing for and dealing with 80–3
   grieving process 82
declawing 70–1
defecating, in the house 67

INDEX

denning site 56
dental hygiene 38
de-sexing (*see also* neutering)
   37, 61
diarrhoea (*see also* FIE) 26
diet 35
   balanced 44, 50, 51
   commercial foods 48, 49, 73
   formulated, therapeutic 51
   bones 53
   dairy 52
   eggs 52
   fast foods 48
   fats and oils 46, 52
   fibre 47, 53
   fish 51
   home-prepared 49, 74
   liver 51
   meat and meat by-products 51
   specialist 49
   therapeutic 49
   vegetables 53
   vitamins 53
dietary needs 48
digestion, aids 46
   fibre 47, 53
digestive upsets *65*, 66
dogs 13, *28*, 36, 44, 57
drowning 117
domestic *4*, 10, 13, 25, *44*
doors, for cats *34*, 67

E

ear injuries 99, 114
   mange mite (*Otodectes cynotis*) 93
eggs 52
electronic pass system 34
Elizabethan collar *113*
endocrine problems 100
energy intake *47*
euthanasia 80, 83
export and import 16

expressions, cat *60*
external parasites 92–3
eye injuries 114
   problems 100–1

F

fabric eating 65–6
fast foods, varieties 48
fats and oils 46, 52
fat-soluble vitamins 53
feeding, area 54
   criteria 55
   kittens 54–5
   nutritional problems 55
feet, vibration detectors 60
Felidae – *see* wild cats
feline cancer 36, 90
   infectious anaemia (FIA) 95
Felis lumensis 10
   manul 14
   silvestris 10
   silvestris ornate 11
   silvestris, silvestris 10, 11
feral 14, 57
FIE (infectious enteritis) 90
fibre 47, 53
first-aid kit, basic 110
fish 51
flea control 37
   powders *92*
   treatment 23, 75
flehming – *see* gape response
flu 89
flukes 95
food availability 49, 56
   calorific value 49
   cost and convenience 48, 49
   nutritional profile 48
   professional formulae 49
   meat and meat by-products 51
fractures 115
frostbite 118

fur, broken open 76
furniture-scratching *69*

G

gape response 57–8
GCCF (Governing Council of the Cat Fancy) 18
genes *14*, 15
   hybrid vigour 25
gestation period 73
gingivitis, gum inflammation 76
glands, anal 58
goldfish *28*, 29
grains 53
grass-eating 46
grieving process 82–3
grooming 30–1, 38–40
   basic equipment 39–40

H

hair or skin problems, detecting 39
hearing, acute 60
heart murmurs *89*
heartworms 95
hookworms 94
house soiling 66
household poisons 119, 120
hunting 68
   instinct *4*, *18*, 28
   skills, inherited *68*
hybrid vigour 25
hyperthyroidism 71

I

immune system 86
immunity 87–8
infections, incubation 26
   spreading to humans 35
   upper respiratory tract 58
insect stings 118
integration of new pets 27, 80
   bird 28

dog 28
internal parasites 93–5
intestinal problems 101–2

J
jumping, dealing with/preventing 68

K
kittens 5, 6, *12*, 13, *19*, *23*, 24, 25, *26*, 34, 37, *52*, *56*, 57, *61*, 71, *90*
   active immunity 88
   antibodies 86
   basic house rules 36
   development of senses 76
   deworming 75
   mental development 25
   milk teeth 38
   physical development 25
   safety 35
   small prey 58
   weaning 75

L
laser therapy *87*
leg and paw injuries 114
leukaemia (*see* feline cancer)
lice 93
life expectancy 76
lions *10*, 13, *33*
liquid medicine, administering *91*, 112
litter tray 36, 64, *66*
   preventing soiling 66–7
   litter, types of 31, 66
liver *48*, 51
   problems 102–3
lordosis *72*
lungworms 95

M
mating sequence *73*
mature cat, caring for 79

deep sleep 76
feeding and drinking pattern 76
arthritis 77
bladder control, loss of 78
blindness 78
blood samples 76
deafness 78
digestive problems 77
final days 80
grooming 76
ill health, early signs 96
lameness *97*
osteoarthritis 77
protein adjustment 76
senility, signs of 79
warmth and comfort *79*
weight loss 77
milk, lactose-free 46
   low-lactose 52
minerals and trace elements 46
mites 93
mixed breed *4*, *72*
motion sickness 41, 42
mouth and oesophagus problems 103–4
   injuries 114

N
neotony 13
neutering 25, 27, 37
nictitating membrane 26
night vision 59
non-pedigree 15, 24, 25
nutritional problems, causes 55

O
obesity *47*
oestrus 37
old cat (*see* mature cat)
Otocolobus manul 15
overfeeding 55

P
pain, headache 97
   internal 97
   spinal 97
   your cat's reactions 96–7
pancreatic problems 102–3
parasites, external 42, 92–3
   internal 93–5
pedigree 18, 25, 24, 27, 40, *72*, *73*
pet passports 42, 43
PETS travel scheme, participating countries 42
poison 119, 120–1
   toads and lizards 118
pregnancy, *see* queen
problems (health)
   blood and circulatory 98
   ear 99
   endocrine 100
   intestinal 101–2
   liver, spleen and pancreatic 102–3
   mouth and oesophagus 103–4
   nervous system 104–5
   reproductive in the queen 105
   respiratory system 105–7
   skeletal joint and muscle 107–8
   skin 108
   stomach 109
   urinary 109
protein, animal, plant 46, 50
   digestibility 46
purring 60

Q
quarantine 42, 43
queens 47, *73*, 75, 105

R
rabies 90–91
   booster vaccination 43
   carriers 90

INDEX

prevention of 91–2
rabies-free countries, islands 42, 43
registration, new cat 27, *88*
    official for pedigree cats 18
respiratory infections 89–90
    feline calcivirus (FCV) 89
    feline pneumonitis 90
    feline viral rhinotracheitis 89
    system problems 105–7
roundworm (*Toxocara cati*) 35, *93*

S

scratching furniture 69
scratching post 32, *33*
senses 58
separation anxiety 38, 71
skeletal, joint and muscle problems *48*, 107–8
structure 15
smell 58
snake bites 118, *119*
social skill development 21
social system 56–7
soiling, preventing 66–7
spider bites 118
sound, high frequency 60
spaying 37
spleen problems 102–3
spraying 58, 67
startle tactics, while hunting 68, 70
stimulation 32
stomach problems 109
stray cats 22
stress-management 20
stud, book 18
    male 25
supplement, calcium 74
    excessive 50
    formulated 50
    pet food 50

supplementation 51
surrogate companion 34
system, feline 56

T

tabby 5, 15
tablets, administering 111
tail injuries 114
talking 71
taming 12–14
Tapetum cellulosum 59
tapeworm 26, 94
taurine 44
teeth 26
    gums *45*
    permanent 38
teething 38
temperament 14, 25, 26
territory 27, 32, *33*, 35, 36, 56, 64
    marking 58, 67
threadworms 94
ticks 42, 43, 93
toilet training 25, 31, 36
Toxocara cati (roundworm) 35
Toxoplasmosis 35, 95
toys 32
training, basic 37
    simple tricks 38
transport, commercial 42
travel *41, 42*
    cage 41
    diseases 42
    documentation 42
    identification, microchip 42, 43
    overseas *42*
    pet passports 42, 43
    PETS travel scheme 42
    quarantine 42, 43
    regulations 42
TVP (Textured vegetable protein) 49

U

underfeeding 55
United States 16
    breed standards 18
urinary problems 109
urination, in the house 67
urological syndrome, feline (FUS) 37

V

vaccination 23, 24, 25, 36, 42, 73, 88–9
    annual boosters *96*
    certificates 42, 43
    feline leukaemia 36
    pedigree 36
vaccines 86–7
ventilation, adequate 38, 41
vermin control 11
veterinary certificate 42
vision, stereoscopic range 59
vitamins 46, 47
    fat-soluble 46, 53
vocalization 71
vomiting 121
    what to do 119–20
    induced 120

W

water *27*, 35
    daily requirement 44, 54
weaning, *19*, 25
weight loss *50*
whipworms 94
wild cats (*Felidae*) 10, 11, 12, *44*, 45
worms 35, 75, *93*, 94, 95
worship of cats 12
wounds, bite and puncture 112

Y

yowling, dealing with/preventing 71

PICTURE CREDITS

Key to photographers (Copyright rests with the following photographers and/or their agents): A/JC=Anipix/Jan Castrium; AL=Alexis; B=Bios; E & PB=Erwin & Peggy Bauer; JB=Jane Burton; RC=R Cavignaux; C=Cogis; BC=Bruce Coleman; D=Dammon; FN=Foto Natura; G=Gissey; H=Hermeline; JJ=Johnny Johnson; L=Lanceau; KH=Klein-Hubert; RM=Robert Maier; GM=Graham Meadows; O=Okapia; PB=Picture Box; HR=Hans Reinhard; SIL=Struik Image Library; KT=Kim Taylor; ST=Sally-Anne Thompson; PVG=Paul van Gaalen; DVZ=Dries van Zyl; V=Vidal; WP=Warren Photographic.

Key to locations: t=top; a=above; l=left; r=right; b=below. (No abbreviation is given for pages with a single image or where all photographs are by the same photographer.)

Endpapers	BC	27		BC/JB	51	SIL	85	WP/JB
1	WP/JB	28	t l	BC/HR	52	SIL	86	GM
2	BC/JB	28	rt	ST	53	ST	87	GM
3-4	WP/JB	28	a	BC/JV	54	GM	88	A/JC
5 b	WP/JB	29		WP/JB	56 t	BC/HR	89	A/JC
5 t	A/JC	30	t	A/JC	56 a	BC/DVG	90	ST
6-7	WP/JB	30	a	GM	57 a	WP/JB	91	GM
8-9	WP/JB	31		BC/JB	58	GM	92 l	BC/KT
10 t	A/JC	32		WP/JB	59	BC/KT	92 r	KH/FN
10 a	BC/JJ	33	t	GM	60	WP/JB	93	GM
11	GM	33	a	WP/JB	61	WP/JB	94 t	C/L
12 t	GM	34		GM	62	BC/JB	94 b	GM
12 a	WP/JB	35		PB	63 t & a	BC/JB	95 t	WP/JB
13	BC/E & PB	36	t	C/H	64	BC/JB	95 a	GM
14 t	GM	36	a	BC/HR	65	BC/JB	96	GM
14 a	WP/JB	37	t	WP/JB	66	C/D	97	GM
15 tl	BC/JB	37	a	GM	68	BC/JB	110 t	KH/B/FN
15 tr	WP/JB	38		WP/JB	69 l	JC/FN	110 a	SIL
15 a	WP/JB	39		ST	69 r	A/JC	111	A/JC
16	GM	40	l	A/JC	70	PB	113	GM
17 t & a	GM	40	rt	BC/RM	72 t	WP/JB	114	GM
18	C/H	40	ra	WP/JB	72 ar & al	WP/JB	115	KH/O/FN
19	ST	41	t	WP/JB	73 t	BC	116	GM
20	C/H	41	a	C/H	73 a	WP/JB	119	C/G
21	C/H	42		C/V	74	WP/JB	122-123	WP/JB
22 t	ST	43		C/H	76	GM	128	GM
22 a	BC/HR	44		BC/PVG	77	GM		
23 t	GM	45		BC/KT	78	WP/JB	**Cover**	
23 a	C/AL	46		SIL	79	WP/JB	back flap	GM
24 l	GM	47		WP/JB	81	A/JC	back cover	WP/JB
24 r	GM	48	t	SIL	82	WP/JB	front flap	SIL
25 l	GM	48	a	GM	83	WP/JB	front, clockwise from	
25 r	ST	49		SIL	84 t	RC/B/FN	top left WP/JB; WP/JB;	
26	WP/JB	50		ST	84 a	GM	GM; FN/B/KH; C/G; WP/JB	

Ginger as a tiny kitten – a long-time companion of Graham Meadows – to whom *The Cat Owner's Handbook* is dedicated.